D1141193

RULES OF ENGAGEMENT

By the same author:

GOODBYE TO AN OLD FRIEND
FACE ME WHEN YOU WALK AWAY
THE MAN WHO WANTED TOMORROW
THE NOVEMBER MAN
DEAKEN'S WAR

CHARLIE MUFFIN
CLAP HANDS, HERE COMES CHARLIE
CHARLIE MUFFIN U.S.A.
THE INSCRUTABLE CHARLIE MUFFIN
MADRIGAL FOR CHARLIE MUFFIN

RULES OF ENGAGEMENT

Brian Freemantle

C

CENTURY PUBLISHING
LONDON

First published in Great Britain in 1984 by
Century Publishing Co. Ltd,
Portland House, 12–13 Greek Street,
London W1V 5LE

ISBN 0 7126 0459 6

Photoset in Great Britain by
Rowland Phototypesetting Ltd, Bury St Edmunds, Suffolk
and printed by St Edmundsbury Press,
Bury St Edmunds, Suffolk
Bound by Butler & Tanner Ltd, Frome, Somerset

For Philip and Stephanie, with love

Prologue

Soon now, very soon. Minutes, that's all. Only about five babies still there: eight at the most. Going to work; everything was going to work! Marvellous. Christ it's hot. Hotter than hell. Hope Hawkins is all right; doesn't sound good from behind. Not good at all. Bad shape. Jesus, that heat! Poor little bastards. Don't weigh a thing. Like feathers. Hope none die, on the way back. Wouldn't look good, arriving back with dead kids. Shouldn't worry. Going to work. Why's he stopped in front? Shouldn't stop. Breaks the line. Firing! We're under fire! Dear God no, please no. Don't want to get hurt. To die. Do something! Somebody do something. Run. Anything. Please do something. Running. Thank God we're running. Don't let me get hit: please don't let me get hit. I'll make any promise, do anything, but please God don't let me get hit. Thank you God. Oh dear God thank you . . .

Chapter One

The kids were bored, like kids always are at grown-up events they don't understand, fidgeting and playing restricted games and wanting the ceremony to end, so they could get to the coke and hamburger stalls, but near Hawkins there was one solemn, round-eyed black child, older than his years, attentive by his mother's side, arm protectively around her waist. Hawkins looked beyond, to the outstretched wings of polished black granite, one reaching towards the Washington Monument, the other to the Lincoln commemoration, and wondered where in that relentless, sterile list of Vietnam war dead the kid's father was named. What *were* the boy's years? Maybe thirteen. Probably wouldn't remember the man then. How many others at the dedication ceremony would be able to recollect their fathers or their brothers? Hawkins remembered his own legendary and famous father telling him of being in a storm ditch along Highway One, on the way to Tay Ninh, with babies just old enough to suckle from flap-eared breasts, sufficiently aware of the difference between incoming and outgoing fire to know when to cover their ears and when not to bother but hold the tit instead. He guessed they wouldn't have fathers either. Not many, anyway. Always the innocents, rarely the guilty. Pompous thought, decided Hawkins, blinking away the reflection, knowing he should have accepted the unexpected invitation to come with the Petersons and guiltily now trying to find them, to regain the opportunity. Bloody stupid, to have made the excuse. Idiotic. There hadn't been any need to check his telex or the agency machines at the Press building. Kept himself to two drinks, though. Singles at that: still a good day, so far. Never going to develop into a problem, like before. Never.

He saw the politician at last, on the Lincoln Memorial side, quite a long way from where he stood, in a predictably reserved, roped-off section. There were photographers clustered around, because the speculation was open now, but Peterson appeared unaware of their closeness, straight-backed and properly solemn faced. Peterson's wife was by his side, pretty in a corn-fed, vitamin enriched milk-every-day kind of prettiness which Hawkins had never liked until he met her and which he now liked very much. Which was ridiculous and he knew it, like saying he had to check the office wires when he wanted to stop off at the bar, instead of travelling in their limousine. He decided there was less harm in admiring Eleanor Peterson – or would lust be a better word? – than there was in two drinks, even small ones, so early in the day. Far less harm. Because those two quick snorts had been more important than getting close to a man likely to become the next President of the United States. And that was the sort of risk he'd promised himself never to take again.

The ceremony ended with an abruptness, practically an anti-climax. People stood as if they expected more and were unsure what to do, with only the impatient children anxious to get out into the Mall. An apparently uncoordinated movement started, to walk by the memorial and study the names inscribed into it and Hawkins went with the crowd to get to Peterson. He remained at the rear of the line, glad his height enabled him to keep the Petersons in sight as the walkway dipped. When he got nearer the Mall Hawkins saw the official group was already walking towards their waiting vehicles, Peterson among them. Hawkins concealed the breath deodoriser in the palm of his hand, squeezing it several times into his mouth and then called formally 'Senator! Senator Peterson.'

The politician didn't hear, continuing on and Hawkins called again and this time the man hesitated and turned, smiling in recognition.

'Ray! Didn't think you were going to make it, after all.'

'Important,' Hawkins lied. 'Kept me longer than I expected: sorry I couldn't come with you.'

Although Peterson wasn't yet a declared candidate the machinery existed because no one had any real doubts the man intended to run. Joe Rampallie, the campaign manager in everything but official title, was further down towards the waiting limousines and Peter Elliston, of whose precise function Hawkins was unsure, was already at the open door of one of them.

Eleanor stopped and turned and when she saw Hawkins she said 'Hi!' and gave one of her open smiles. She had brace-straight, even teeth.

They were an impressive couple, Hawkins decided: a presidential package. Peterson was very tall – as tall as he was, which was six-four – and exercise slim, which he wasn't. There was a flush of health about the man's face, too, and Hawkins watched as Peterson brushed back from his forehead in an habitual gesture the shank of flax-hair that the cartoonists were so fond of caricaturing. Eleanor was not as tall but still big for a woman, bright-complexioned and blonde-haired, fittingly dressed in black. Hawkins wondered if the respectful outfit had been her choice or that of the election machine. He was conscious of his out-of-condition paunch against Peterson's slimness. Maybe he'd get in shape soon. Another maybe.

'John said you had to stop by at the office,' she said.

'Always queries from London,' said Hawkins, continuing the lie.

'How are things?' said Eleanor.

'OK,' said Hawkins. He'd returned from the funeral in England two weeks earlier. 'I meant to call you, to thank you for the wreath. It was kind.' Something else put off, he thought.

'Sure you're OK?' she pressed, unconvinced.

'Positive,' he insisted.

'Now that you're back we must lunch together soon,' said Peterson. There was a pause. 'Up on the Hill.'

'I'd like that,' said Hawkins, aware from the qualification that it was to be an official, not social occasion. He'd need invitations like this now. And an in-depth profile on Peterson, in advance of his actual declaration, would make a good comparison piece with the incumbent President. The incumbency gave Nelson Harriman the traditional advantage but during his initial term unemployment had risen to record levels and Harriman had been unlucky with the timing of a gesture to Moscow, accepting a Soviet invitation to a fresh round of arms reduction talks a month before intelligence confirmation that the Russians had developed an improved

SS-20 missile, exposing himself to the accusation of being weakly conciliatory.

Peterson was a formidable contrast, a recognised war hero whose consistent championing of Vietnam veterans was getting him national popularity now the pendulum of public feeling was swinging against the previous rejection. But Peterson wasn't just relying on the past. During the most recent European tour there had been meetings with the British Premier, the West German chancellor and the French President and speeches in every capital insisting upon confronting Russia through strength, not weakness. And the domestic record was good, too, with an equally consistent voting record on welfare. Three days earlier Hawkins had watched the man on *Good Morning America* convincingly put the case for lower interest rates to stimulate the stalled American economy and out-argue a guru monetarist.

Definitely make a good feature, predicting the unseating of the first incumbent President since Jimmy Carter, Hawkins determined. Too good to have risked for a couple of drinks. No problem, though.

From behind them Hawkins saw the campaign manager hurrying back from the waiting cars. Joe Rampallie was a saturnine, unsmiling man who blinked with rapid nervousness from behind glasses that were advertised as being the executive type. They had only met three times but Hawkins didn't think the other man liked him. Rampallie said something that Hawkins didn't hear and Peterson looked uncertainly at Eleanor, who didn't respond. The senator shook his head at Rampallie.

'You must come out to the house again soon, too,' said Eleanor, to Hawkins.

The last time, six months before, Hawkins had gone by himself because by then his father had been too ill. He'd drunk too much scotch before he'd left their own house and continued on highballs in Georgetown and got into a confused, unconvincing argument about US policy in Latin America with a Democratic senator from Iowa who'd lost his temper and accused him of not knowing what he was talking about. Hawkins feared the absence of an invitation since then had been to show disapproval. Or maybe it was because Eleanor guessed how close he'd come, in his drunkenness, at making a pass at her. Thank Christ he hadn't. 'Thank you,' he said. 'That would be very nice.' Next time he'd stay sober, for all sorts of reasons.

'I'll call,' promised the woman.

To Peterson Hawkins said, 'I wondered if the others would be here today?'

'Others?' said Peterson.

'Blair and Patton,' said Hawkins.

Peterson frowned, looking into the crowd as if seeking the other two survivors from the mission. 'I thought I saw Colonel Blair earlier.'

Behind the politician Hawkins saw Rampallie beckon Elliston from the vehicles and enter into an immediate conversation. At once the man hurried away into the crush of people.

'I expect they'd come,' said Hawkins.

'I guess so,' agreed Peterson. 'We haven't kept in close touch: just Christmas cards, stuff like that.'

There was no reason why they should have done, Hawkins supposed. With his father they'd been brought together in Vietnam by an event, not through friendship. Rampallie had another of his soft-voiced exchanges with the politician and Peterson shook his head. 'I don't think that's a very good idea,' said the senator.

'It can't hurt,' said the campaign manager, his voice reaching Hawkins.

'I don't like it,' insisted Peterson.

'The organisers won't mind; it gets them coverage.'

'They've got that anyway,' said Peterson. 'It's cheap.'

'Looks like it's fixed,' announced Rampallie and from the direction of the monument Hawkins saw the slim, urgent staffman returning. There were three people with him, a soldier whose chest was a technicolour of decorations, a plain-suited man and a woman, dressed like Eleanor, severely in black.

'Don't do anything like this again,' Peterson said to Rampallie, as he turned to meet the group. To Hawkins he said, in introduction, 'This is Colonel Elliott Blair . . . Eric Patton . . .' He paused, smiling without recognition, at the woman.

'Sharon Bartel,' she said. 'My husband was Major Bartel.'

Peterson cupped both hands around that of the woman and said, 'Of course. It's good to see you after so long, Mrs Bartel.'

Charles Bartel had been the junior ranking co-pilot of the helicopter, recalled Hawkins, one of the eight who didn't make it. Francis Forest had been the later condemned Green Beret Colonel in command of the mission and two other Green Berets, Frank Lewis and Paul Marne had died with him. So had Howard Chaffeskie and James McCloud, the sidegunners. And the civilians, Harvey Lind, the CBS cameraman whose film deservedly won him a posthumous award and John Vine, the CBS reporter for whom the excursion into Chau Phu had been the final assignment.

As the hurriedly assembled group went through the ritual of introductions, Hawkins was aware of Rampallie summoning the remaining photographers and of a sudden scurry towards them. Peterson was right; it was cheap.

'I'm sorry about this,' apologised Peterson. 'Really I am.'

They stood around with the discomfort of strangers unsure how to react to one another, each waiting for the other to initiate a conversation.

It was difficult to calculate the number of decorations Blair carried but Hawkins guessed it had to be more than fifty. He isolated the Silver and Bronze Stars and more than one Purple Heart and gave up. Like counting fireworks, he thought. Blair looked like a model for a recruitment poster, hair shorn to the skin high above his ears, stiff-backed, uniform immaculately uncreased, shoes gleaming. The hand contact between them was brief but Hawkins' impression was of physical hardness; knocking on Blair's chest would be like knocking on wood.

Eric Patton was quite different, a tall, polished-faced man with retreating hair. He bulged with plumpness around a waistline where one-time muscle had become fat through neglect or indulgence. His attitude was of an eagerness to please and Hawkins realised that the hesitant smile seemed always waiting, in readiness.

The widow of Charles Bartel was a petite, quietly still woman whom Hawkins guessed to be only three or four years older than Eleanor Peterson, although the age difference appeared greater. Eleanor Peterson had the beauty of someone protected throughout her existence: close up there was a tightness about Sharon Bartel's face and there were pinch-lines in the corner of her eyes. Happy beauty against sad beauty, Hawkins thought.

It was the practised politician and his wife who made the conversation.

'I'm sorry,' Eleanor said to the other woman. 'It must be a difficult moment for you.'

'There have been quite a few,' said Sharon Bartel. 'I've become used to them.'

'What do you think of the monument?' Peterson asked generally.

'It's not enough,' said the widow, positively. 'I've read all about the supposed artistic merits but I don't think it's sufficient.'

'There's supposed to be a sculpture planned,' said Hawkins, entering the conversation.

'And provision for a flag,' said Sharon. 'I still don't think it'll be a sufficient improvement.'

'Maybe it's something we have to get used to,' suggested Patton. The hopeful smile flickered on and off, like an uncertain light.

'Maybe,' said Peterson. 'What do you think, Colonel?'

'Appropriate,' said Blair, curtly.

Hawkins guessed that any other stance than that of almost upright attention would be difficult for the soldier. His father had told him that in Vietnam the Special Services were the military elite, the supermen: whenever Green Beret special forces entered a bar or restaurant, ordinary soldiers had risen to offer their seats. Blair looked as if he would have taken such homage for granted.

'I met Harvey Lind's wife at the beginning of the ceremony,' said Sharon, looking back into the crowd in apparent search.

To Rampallie Peterson said, 'I should write to her: Vine's wife, too, if he had one. Remind me about that.'

Patton smiled at Hawkins and said, 'Where's your father?'

'He died, a month ago,' said Hawkins. 'I managed to bring him to see the preparation before he became too ill.'

'I'm sorry,' said Patton automatically.

'He was a good writer,' said Sharon Bartel. 'A sympathetic man.'

'Yes,' said Hawkins, greeting the familiar praise. 'He was.'

'He contacted me afterwards,' she said. 'Wrote me a very nice letter.'

'After Vietnam he was posted here, as the Washington correspondent,' said Hawkins. 'I succeeded him two years ago, but he stayed on. We lived together.'

'Quite unusual, son following father like that?' offered Blair. He had a flat, unmodulated voice.

'I suppose it was,' agreed Hawkins. Unusual wasn't the polite word in the London office, Hawkins knew. Just as he knew there was some justification for the sniping. He wouldn't have got directly into Fleet Street from Cambridge without his father's influence. Or been kept on, when the drinking had become established and the mistakes had started, bad enough for two libel writs. Or – from that mess – got the Washington job, a prestige and reward posting on a Sunday newspaper as politically respected as theirs. Or managed to sustain it, hopefully appearing from the distance of London to have controlled the drinking and created his own contacts which were, in fact, a legacy of his father's influence and respect in the American capital. Or . . . Abruptly Hawkins shut off the litany going through his mind. All over now, he thought. Now he was on his own. Very much on his own.

A lot of executives had attended the funeral and later the memorial service in St Bride's. It was at the memorial service that he realised there were some of them who suspected what had happened in America. And were waiting for the proof, like vultures in the nearest tree. Hawkins swallowed, his throat moving, wishing he had a drink. Very much alone, he thought again.

It was Eleanor Peterson who brought the awkward encounter to an end. Indicating their waiting car she said, 'Can we give anybody a lift anywhere?'

'I've got a car, thank you,' said Patton.

'No thank you,' said Blair. 'I've got transport too.'

'I'll walk,' said Hawkins. 'Fourteenth Street is close enough.' And the bar on Virginia Avenue was even closer.

'Don't forget what I said about keeping in touch,' Peterson reminded, moving with Rampallie towards the cars at last.

'I won't,' said Hawkins.

Peterson's car was an executive Cadillac, with secretarial jump-seats facing the rear seat passengers and small courtesy tables that swung outwards from the door supports, as a writing platform between them. Rampallie sat directly opposite the senator, with Elliston facing Eleanor Peterson.

'What's the schedule?' demanded Peterson.

'Nothing until the *Johnny Carson Show*,' said the campaign manager, diary opened before him. 'That's a recording, of course, so there's time for the dinner. You're speaking in support of Abe Schuster, who's guaranteed the mayoralty, so we're backing a winner.'

'Work some criticism into the speech of the New York Irish support of the IRA against Britain,' advised Eleanor.

'There's a strong Irish vote in New York,' reminded Rampallie.

'Which is why it's the platform to make the criticism,' insisted Peterson. 'It's an undertaking I gave in London. I think I'll raise it on the television show, as well.'

Rampallie gave a shrug of acceptance.

'Whose idea was it to bring Patton and Blair and the woman over?' demanded Eleanor.

'Mine,' admitted Rampallie at once.

'It stank,' said Eleanor.

'I didn't mean to cause any upset,' apologised the man. 'It seemed a natural, that's all.'

'It could be misinterpreted,' said Peterson. 'I don't want the criticism that I'm making tacky capital out of it.' To Elliston the politician said, 'Don't forget those letters to the families of the television team. And the others, too: Lewis and Marne.'

To Eleanor he said, 'I thought Ray looked flaky.'

'I thought he looked all right,' said the woman.

'You always have,' said her husband. 'His breath smelled of freshener.'

'Maybe it was social politeness.'

'Maybe it was scotch.' The huge car turned off the Washington Circle towards Georgetown and Eleanor said, 'It was a good idea to invite him up to the Hill.'

'I thought so,' said Peterson. 'His newspaper is the most influential Sunday in England. Friendly, too. Met their proprietor in London. A lord called Doondale.'

'Ray's got a hard act to follow, in his father,' said the woman.

At that moment Hawkins was sitting on a bar stool on Virginia Avenue, both hands gratefully around the glass, staring down into the drink. Just a couple, he thought; no more. No problem. Not any more.

It was after lunchtime when Hawkins got back to Fourteenth Street and his office in the National Press Building. The message on his incoming telex machine asked him to telephone London and when he did he was told to make arrangements to fly back to England immediately. He decided he needed a drink.

Chapter Two

Hawkins had a couple of drinks on the aircraft and felt satisfied with himself because he could have ordered more but didn't, just wine with the meal, and he'd had nothing since he'd got to the Waldorf, which was two hours ago so it was pretty good. Damned good, in the circumstances. Because there couldn't be any doubt what the circumstances were. There must have been more vultures than he imagined. Still quick. Little point, really, in letting him return to Washington after the funeral. But newspapers were never logical. Couldn't be fired, Hawkins thought, seeking self-encouragement: there mightn't be any confidence in him and they might have called him back for reassignment but they couldn't fire him because there wasn't any reason, not recent enough to provide the justification anyway. He picked up the room service list, looked at the drinks prices and put it down again on the table. What would they offer him? General features, maybe: that was usually the graveyard where elephants who confused their tricks were sent to die. Maybe a title – special writer was the favourite – in an effort to show he wasn't being dumped, which would be confirmation to everyone that he had because they always invented special titles to try to cover it. Maybe they'd just come out into the open and say that it was all over now that his father was dead and that there was a place on the foreign desk and he could take it or leave it. He picked up the drinks tariff and decided the mark-up was high, carrying the sheet with him to the window: below the rush-hour traffic was clogged around the island dominated by the BBC building.

So what would he do, take it or leave it? Christ, how

wonderful it would be to tell them to stick it up their ass and walk out and pick up another foreign posting on a rival Sunday, the *Observer* or the *Sunday Times* perhaps and write the hell out of whoever they put in against him! Wonderful and about as practical as winning the Pulitzer Prize for Journalism, which was restricted exclusively to American correspondents except for the one unprecedented special occasion when they created a category for his father. Hawkins shivered, as he'd shivered at the funeral and for the same reason, a physical reaction to coldness. He was, he realised, cold and exposed, locked out and alone.

He turned away from the window and the jammed traffic, determinedly replacing the bar list once more. Then he shifted it again, folding back the cover to create a stand and placing it upright against the side of the mirror, so the drinks confronted him in constant challenge. He sat at the narrow desk and stared at the listing with the intensity with which he'd tried to absorb the text books and the lecture notes, remembering the determination to get the First Class degree his father had so much wanted him to gain. There hadn't been any rebuke or criticism at the failure: there hadn't been the need, for Hawkins fully to realise the old man's disappointment. One among so many, thought Hawkins.

He looked away from the tantalising list, directly into the mirror alongside. A haircut would help, he saw. He'd have one before the meeting with the editor and perhaps let the barber grease it, although normally he didn't bother: at the moment it curled lank to his collar, any parting lost in the disorder like a forgotten path beneath out-of-control undergrowth. Yes, definitely a haircut. He strained forward, blinking. Eyes not too bad; actually very good, hardly bloodshot at all and what there was easily understandable from jet-lag. Hawkins was prominent-nosed, a characteristic inherited from his father whom he strongly resembled and he felt up with his fingers, moving it exploratively left and then right. Pretty good again: right side unmarked and only the vaguest tracing of broken veins on the left, hardly noticeable at all unless you were positively looking, right up close like he was now. Skin wasn't good, though: waxy and dry-scaled, some actually shed over his jacket collar, a white debris. Have to rub some Vaseline into his cheeks, before the meeting. Some eye drops, too, to get rid of what little redness there was. Could have been worse; a lot worse.

So what would he do, if the best offer was a telephone-answering desk job? he demanded of himself. Quit? It wasn't admitted to probate yet – wouldn't be, for a few months – but there was probably thirty to forty thousand in his father's estate, because he'd been a careful man, despite the apparently adventurous life. And the Washington house had gone up in value since they'd bought it: could be another eight to ten thousand profit from the sale of that. Enough for three years: four if he were careful. Quit to do what? The great unwritten novel, the supposed ambition of all journalists? Hawkins sniggered at the idea. He didn't have a book in him: hadn't ever considered it. What then? He didn't know. And with his father gone, there wasn't anyone to ask. Was there *really* a question anyway? Hawkins knew what he would do – like he'd always known what he would do – whatever he was offered.

Take it. Hope for a bullshit title or some empty cosmetic wrapping and take it. OK, so he'd done it through his father but no one really knew that, so on the surface he'd held the Washington posting down for two years with more than enough successes and hardly any failures, and they couldn't take that record away from him. Unlikely, really, that the offer would be a desk job. Take whatever it was then: take whatever it was and prove himself as good as his father. Hawkins stopped at the thought, curious at it. Always the intention but never the chance, he reflected. It was wrong to blame his father – ungrateful because everything the man had done had been to help and encourage – but his own entry into journalism had coincided with his father's permanent basing, first in London and then Washington, so throughout his father had been behind him, like a shadow, showing the way and opening the doors and pointing out the proper bridges. There had been times – a lot of times – when he'd felt claustrophobic, enmeshed and restricted. Now the shadow wasn't there any more he realised just how much he had depended upon its shade.

The bar list drew his eye, like a coloured banner. Two pounds for scotch. High compared to the press club prices perhaps but not against the Washington Hotel or the Mayflower or the Jefferson or the Four Seasons. One wouldn't hurt. Thousands – millions – of people had a nightcap: healthier than sleeping pills, in fact. Wouldn't be like going down to the bar, where he could nod for another. Order just one from room service: limit himself. Hawkins pushed himself up from the desk, a rejecting motion, undressed, got into bed and lay propped up, looking at the beckoning tariff. Didn't need it: *wouldn't* need it.

Hawkins left the hotel early and because it was so close decided to walk to Fleet Street. He had his hair cut in Chancery Lane, bought some eye drops and Vaseline in an adjoining chemists and because he was early for the appointment and because he'd proven his resistance the previous night allowed himself a reward and went into El Vino. He'd used it a lot before following his father to America but the barman was new and there was no one else he recognised. After one drink he had another, not because he desired a second but as a test to confirm last night's strength that he could hold back from going on to a third, which he succeeded – just – in doing. In the downstairs toilet he applied the eye-drops and lightly greased his face and used the mouth spray, and he squirted the spray again in the lift taking him up to the editorial floor when he reached the office. It had been redesigned since he worked there and he was glad; a separating corridor meant he could reach the editor's room without being visible to the open work area. He didn't want to run a gamut.

David Wilsher was an ebullient, enthusiastic man very aware of himself and of his place in a village as small as Fleet Street. He'd been appointed editor when the paper was ailing and facing closure, changed it to a tabloid format and surpassed any previous circulation figure. In so doing he became a public figure, a frequent television face and a feature subject in colour magazines and journals. He worked at being a recognisable character: he sometimes rode a thousand c.c motorcycle to the office instead of a Ferrari, enjoyed hang-gliding and crewed 12 metre yachts well enough to be short-listed for the America's Cup team. Wilsher was a proponent of personalised journalism, hence the association with Hawkins' father had been so long and so successful, because that was the way his father wrote. In trying to decide between admirers and detractors Hawkins decided that in Wilsher he had a friend. Objectively he recognised that it was a friendship that would last only as long as it did not endanger Wilsher, who enjoyed his reputation and wouldn't allow it to be put at risk. Which was what Hawkins guessed the man now feared.

There was a pumping handshake and the offer of a drink, which Hawkins pointedly refused. They had coffee instead.

'Didn't expect to be back so soon,' said Hawkins. There wasn't really any point in patting pleasantries back and forth across the net, he decided.

'Almost came out to see you,' said Wilsher, refusing to be hurried. 'Too long since I was last in America. Problem with sitting in this chair is that you're expected to do it all the time.'

'Washington is pretty cold at this time of the year,' said Hawkins. Wilsher was in charge: let him dictate the pace.

'Think we'd better get Harry in on this,' said Wilsher, pressing the summons button on the intercom. Hawkins shifted uncomfortably. Harry Jones had the title of assistant editor but his predominant responsibility remained the foreign department. Jones had been foreign editor at the time of Hawkins' surprise appointment to Washington, a decision made between his father, the editor and the proprietor without any reference to Jones, who coveted – and expected – the posting himself. Instead he got the consolation promotion but during the time Hawkins had been in the American capital Jones had been his biggest critic. A patient – and now rewarded – witness for the execution, Hawkins thought.

Jones entered the office in his shirt sleeves, a bustling, bespectacled man, appearing surprised to see Hawkins in the room.

Wilsher said, 'Want you to be in on this, Harry. Should have told you earlier.'

Why didn't Jones already know? Hawkins thought at once. The man would have been the architect of any move against him.

The assistant editor nodded a greeting to Hawkins and said, 'What is it?'

The editor grinned between the two men and Hawkins guessed he'd had his teeth capped. 'An idea from the Almighty,' said Wilsher, raising his eyes theatrically upwards. 'Doondale himself. At least this time it's a bloody sight better than it normally is.'

Hawkins wished he hadn't had the drinks; even just two. It wasn't going as he expected and he was finding it confusing. 'What?' he said.

'Something good,' said Wilsher confidently. 'Your father was the star of this paper, a legend for longer than most people can remember. Read him myself when I was a kid: fantasised . . .' The man paused, to convey that although there'd been youthful admiration, he now considered himself to have attained greater heights. He continued, 'Doondale is commissioning a bust, for the front hall. There'll be a plaque, too. And an unveiling ceremony. The whole thing.'

'That's very flattering,' said Hawkins. Surely he hadn't been brought all the way back from Washington to be told this! Possibly, if Doondale had decreed it. Did it mean he was safe, for the time being?

'But that's only half the idea,' Wilsher went on. 'The other is that there should be a biography. We'll serialise it, of course. Give us the chance to reprint some of the best material he wrote. Doondale owns half Bedford Square, so a publisher won't be difficult to find, especially with agreed serialisation.'

'Who . . .?' started Hawkins and then stopped, knowing.

'Who else!' said Wilsher, laughing at the other man's surprised realisation.

Hawkins was conscious of the intense, rigid attention of Jones but other realisations crowded in upon him, pushing that impression aside. He *was* safe! Not just safe; fireproof. This paper – like every other paper with an individual, idiosyncratic owner – *did* respond to personal demands as if they were Holy Writ. And it was all going to be so easy! They didn't know it but it was all going to be so ridiculously, bloody easy! He felt a feeling better than any drunkenness: it was a euphoria, a numbing, tingling lightheaded wonderful euphoria. Knowing he should respond Hawkins fell back on to a cliché and said, 'That's going to be . . . ah . . . a great challenge.'

'Think you can meet it?' demanded Jones at once, seeking a weakness.

'Of course I can meet it,' said Hawkins, recovering further.

'Everything will be available here, of course . . . files, stuff like that,' assured Wilsher.

There wasn't the confusion of alcohol now. Hawkins' mind was itemising the advantages and benefits. 'I'd like to do it from Washington,' he announced.

Wilsher shook his head doubtfully. 'I was going to suggest to Harry here that we give you a sabbatical . . .' Wilsher thrust out his hand, in a reassuring gesture. 'Washington's yours. No question of that. But this thing is important: Doondale important. You could go back to Washington afterwards.'

'I can do it from Washington,' insisted Hawkins. Because in the basement of the house in Maryland Avenue were the records and the memorabilia and the documents amassed by his father right from the time of the Second World War.

'It wouldn't be a problem to put somebody else into Washington,' persisted Jones, relentlessly.

It would be for me, you bastard, thought Hawkins. He said, 'Neither will it be a problem for me to do both things. There's the forthcoming presidential election. I should be there for that.'

'I respect your professionalism,' said Wilsher. 'But with the proprietor involved normal considerations can be adjusted.'

'If I thought I was endangering the project I wouldn't suggest it,' said Hawkins, probing his new strength.

'I'm not sure,' said Jones.

Seeing the door closing against him, Hawkins said, 'The moment I think either is threatened . . . the Washington coverage or the book . . . I'll say so.'

'I think there should be a monitor on the book,' said Jones. 'We should be kept informed of how it's going . . . outlines, chapter plans, everything like that. There's going to be a lot of pressure from upstairs.'

'I agree,' said Wilsher. To Hawkins he said, 'OK?'

'OK,' said Hawkins. What he saw as his salvation – temporarily at least – Jones saw as the vehicle for his destruction. Fuck the man.

'We can judge the Washington coverage easily enough by what appears in the paper,' said Jones.

'I'd prefer it to be judged by what is filed,' said Hawkins, unwilling to give the man any advantage. 'There are too many things that can affect what appears in the paper.'

'This has got to be bloody good,' insisted Wilsher. 'Apart from the internal interest it's got a lot of mileage for the newspaper . . .' The man allowed a gap and then added, 'Could have a lot of mileage for you, too.'

'I appreciate that,' said Hawkins. He would – later – he was sure. But not now.

'Anything you want, just shout,' promised Wilsher. 'Harry will be liaison, OK?'

No, thought Hawkins. Unable to argue, he said, 'That's fine.'

'Let's make sure this thing works: works like hell,' said Wilsher, in his enthusiastic, follow-me-out-of-the-trenches voice.

That night Hawkins did order from room service and sat in the same chair as before, gazing into the mirror at the same reflection: the carefully arranged hair was overgrown again and his eyes were redder. He giggled, raising the glass to his own image in a solitary toast.

'Here's to success,' he said to himself. It had been stupid worrying himself into the state of depression that he had: absolutely bloody stupid. He had nothing to worry about: nothing at all. It was all going to be so easy.

Peterson took one of the best suites at the Four Seasons, on the corner, so from the panoramic windows there was a view up the Potomac and over Georgetown itself. The chairman of the national committee was there, together with the national treasurer and the other two were regarded as the most influential of the organisation.

'It can't have been pleasant, having to go into such detail,' apologised the chairman. 'It's important we don't make any mistakes, you understand?'

'Of course I understand,' said Peterson.

'You must understand also that this is completely unofficial: this meeting never took place.'

'I know,' said Peterson.

'You've a fine record, sir,' said the chairman. 'A very fine record. We're grateful for your complete openness, about everything.'

'I recognised the purpose,' said Peterson, intentionally subdued.

'When are you going to declare?'

'Soon now,' said Peterson.

'We won't be able to come out officially, not yet,' said the man.

'I want you to know that I deeply appreciate your support,' said Peterson. 'Deeply appreciate it.'

'Winning's the important thing, senator,' said the national chairman. 'After checking everything as thoroughly as we have tonight my colleagues and I have the feeling that you're a winner.'

'I hope so,' said Peterson, sustaining the modesty.

'Hope hasn't got anything to do with it,' said the chairman. 'Our job is to make it happen.'

'What about that!' Peterson said later, when he recounted the meeting to Eleanor.

'Like the man said, all you've got to do now is win.'

'How much do you want to be First Lady!' he demanded, feet off the ground in his enthusiasm.

It was several moments before Eleanor replied. Then she said, 'I'm not sure.'

Chapter Three

His father already had the wooden clapboard on Maryland Avenue when Hawkins arrived in Washington, to take over the posting. There was a view of the Capitol dome from the verandahed upstairs rooms, reached by a wide central staircase which began from a polished wood hallway with a fireplace that was necessary in the winter because it was an old property with inadequate central heating, something Hawkins found strange that his father was able to accept after spending so much of his life in tropical climates. It was an expansive, sprawling place – which was why Hawkins had moved in with the man – with separate dining and drawing rooms and an L-shaped kitchen big enough to eat in. There was a study-library, too, in which he and his father spent most of their time, a cluttered place of dishevelled desks and overflowing shelves that the cleaning woman had instructions not to disturb. And the basement running the length and width of the property. His father had been a meticulous – almost fanatical – keeper of records and they were all stored in the huge downstairs room, in carefully dated boxes and cases. There was volume after volume of hard-backed album-sized books in which were preserved and annotated every story upon which his father had ever been assigned, weekly if not daily diaries of more than thirty years of being at the centre of every newsworthy war theatre. And there was a great deal of personal memorabilia, signed letters and photographs from statesmen – some successful, others failures – yellowing, stained notebooks, even rough copies of the stories that made world headlines.

Hawkins admired and respected – and yes, was in awe – of his father but he'd never known the man; not known him like sons who could remember outings and intimate anecdotes or even growing-up arguments. During his childhood and youth and even at university his father had been the famous man on the front page, never the person he went home to during exeats and vacations, which were always spent with sympathetic parents of other pupils. He had a lot in common with the kids at the dedication ceremony, he thought.

He'd tried to know him, in the final years in Washington, during the long rambling discussions upstairs in the study, usually at the collapsing fire, bottom-of-the-bottle time of the day. To learn to love him even. But never managed it, not as he felt he should have managed it. Perhaps actually writing about the man would provide another chance; perhaps when he'd finished the manuscript he'd know something like love. Not that he trusted himself with an emotion like love, not since Jane. He was surprised at her sudden entry into his mind. He hadn't thought of her for a very long time. But he hadn't felt this lonely for a very long time.

Easy though he knew it was going to be, Hawkins didn't minimise the importance. He'd been frightened in London, confronting for the first time how vulnerable he was. He wanted an opportunity to prove himself and this was it, not just with the book but with the undertaking from Wilsher that his job in Washington was safe. So it was important not to fuck it up, like he'd fucked up so much else. Which meant watching the booze. Accordingly he carried only beer with him to the basement and only a six pack at that, placing it carefully beside the small, specially arranged desk he had brought down from the study, together with the anglepoise light.

Begin at the beginning, he decided logically. Asia then, ironically where it started and where his father's front line career ended almost thirty years later. His father was in a commemorative picture of the Japanese surrender to Mountbatten in 1945 and the strap across the front page headline described him as 'the right man in the right place'. He'd only been twenty-five years old then, campaign-thin because he always moved with the troops, never staying behind for the propaganda hand-outs with the other correspondents in the comparative comfort and guaranteed safety of command headquarters. That had been a favourite study boast, from a man entitled to boast – 'to tell the truth a reporter has to see, for himself. Hacks take it second-hand, writing propaganda. Never be a hack.'

It was an axiom for the book and one the man always followed, right up to Vietnam, Hawkins realised: important not to forget it when he began the narrative.

The man had stayed in the Far East, covering the French effort to recover Vietnam in 1946 and then the emerging independence of India and the conflict between Moslem and Hindu that followed. There was Malaya and Korea – with brief return visits to Vietnam and the French – after that. With the insistence of always being in the field Edward Hawkins was making legends now: and having labels attached. 'The 30,000 miles a year man' was one at which Hawkins smiled, turning the scrapbook at random. 'The World's foremost war correspondent' was another.

Fleet Street exaggeration, Hawkins supposed, but with some justification: the coverage wasn't restricted to Britain now but was being distributed by the newspaper's syndication service and appearing in America and Canada and Australia and even in Europe, in translation.

It was in Korea that his father got married, to the twenty-two year old English nurse seconded to the base hospital in Seoul. There was a picture of the ceremony being conducted by an army chaplain in what appeared to be a tent and Hawkins made an entry into his index, considering it for inclusion in the book. His father had told him how the promised honeymoon in Tokyo never materialised because of another war and another battlefield, at a place then known as French Indo-China.

Hawkins leaned back in the chair, opening the first of the beers. It couldn't have been easy for his mother. Camp following, she called it, according to his father. Nominally she'd lived in Hong Kong, with trips to Tokyo or Bangkok or Manila. It was in the Philippines she contracted the malaria that had killed her, so young that Hawkins had difficulty in remembering her. Certainly his recollection wasn't of the woman whose pictures he was staring down at now: to Hawkins his mother had always been wasted and exhausted, never as vibrant and alive as she appeared here.

There was a picture of his mother and father with a child between them he presumed was himself. Inscribed on the back, in his father's neat, precise handwriting, were the words 'Off to Saigon'.

His father's last visit to Vietnam, to record the desperate French attempts to retain its former colony, lasted right up to the ignominy of Dien Bien Phu.

Predictably his father had actually been at Dien Bien Phu, able to report the initial attacks in the early March of 1954. Even when the Viet Minh destroyed the first airfield and badly damaged the second he stayed on, yet again the right man – journalistically – in the right place. He left, one of the last civilians out, on the thirtieth day of the seige.

Edward Hawkins moved from Asia, after Vietnam. From their study conversations Hawkins knew that he and his mother had returned to England and a rented cottage in Sussex while his father had stayed with the French, throughout the Algerian war, by now the acknowledged doyen of foreign correspondents, the first man independent enough and with sufficient political acumen to write the unthinkable, that France would have to surrender that colony so soon after quitting another further east.

There was the Congo after that, not just with searing eye-witness accounts of the atrocities but with exclusive reporting of the American part, through the Central Intelligence Agency, in the installation into power of Mobutu. After the Congo came Kenya, during the Mau Mau uprising, more atrocities again. And Cyprus, the tug-of-war island between Greece and Turkey.

All carefully recorded in the books that had gradually grown around Hawkins like some protective wall as he had flicked through them. Secessionist wars, every one, Hawkins realised. And already from his brief scrutiny he knew that his father had always recognised them for what they properly were and anticipated the eventual conclusion; another point to go on to the filing cards, for inclusion.

Hawkins idly opened the last volume.

His father was a confirmed legend by the time of the changed Vietnam war – the American involvement – the man by whom statesmen actually sought to be interviewed. Like the practical journalist he was the man had used the advantage to the fullest. As early as 1967 there was the World exclusive interview with Ho Chi Minh in which the North Vietnamese leader insisted on the ultimate outcome eight years later. Exclusive, too, was the interview with America's President Johnson. And Nixon. And not only with Diem and Thieu, the consecutive South Vietnamese Presidents installed by Washington, but with Vo Nguyen Giap, the Communist troop commander who drove them from power as he'd driven out the French, two decades earlier.

From their brief time together Hawkins remembered his father's descriptions of the prestige interviews. 'Sunday suit best' was what the man called them. But always, Hawkins realised, going through the final book, his father had returned to what he was, the eye-witness reporter on the ground.

Which is what he had been on the day of the mission in April, 1975, just one week short of the final collapse of Saigon and the South. Hawkins turned to the story, one of the final pieces in the final album; only the account of the later, official enquiry followed it.

His father's report of the mission and the build-up to it had been given massive coverage, the entire front page allocated to it and then turning to two inside pages, with a third devoted to the photographs of fifteen orphans who had been rescued. There were photographs, too, of Elliott Blair, only a major in the Green Berets then, and of Eric Patton, looking stunned and shocked beside his shell-and-bullet pocked helicopter. There were several pictures of John Peterson, of course, one of the youngest senators in Congress then but already a media figure for his fearless attacks from the Senate floor upon the peace settlement – 'peace with too little honour' was his phrase – which meant the absolute abandonment of South Vietnam.

Peterson's proclaimed purpose in going to Vietnam had been to expose that abandonment. Brilliantly, he chose to show the plight of orphans, not just ethnic Vietnamese whose parents had been destroyed by the war but illegitimate children born to Vietnamese women from American fathers and threatened with Vietnamese rejection because of their bastardy.

The later enquiry made several criticisms, one of which was the wisdom of making available to Peterson in the first place a helicopter and guard unit for the flight into Chau Phu but they acknowledged, too, the panic and confusion of the time.

Colonel Forest had been commander of the Green Beret unit, with Blair his Second-in-Command. Frank Lewis and Paul Marne were the other two Green Berets. Charles Bartel, whom everyone called Chuck, was the pilot, with Patton the co-pilot and Howard Chaffeskie and James McCloud side-gunners manning the miniguns on the UH-1D helicopters.

Hawkins knew from the conversations with his father how the press coverage had been arranged, a coverage which Peterson at first refused because his information was of an abandoned American-financed orphanage at Chau Phu and he wanted maximum space available on the helicopter to bring the children out on the return flight. Only the concerted pressure from assembled journalists made him change his mind and then with the minimum of cooperation. The senator insisted only three journalists would be allowed, out of the fifty demanding access. He personally chose Edward Hawkins as the pool correspondent for writing journalists and CBS cameraman Harvey Lind and reporter John Vine to provide television films.

Peterson's information was correct. There was an orphanage at deceptively quiet Chau Phu. And inside, huddled without food or water or care, were nineteen mixed-birth babies left by a Vietnamese staff who feared retribution from their communist conquerors if they were discovered working for an American organisation.

There would have been a danger from anyone of lesser experience or ability becoming overwhelmed by what happened but Hawkins professionally recognised his father's account to be a superb example of restrained, objective reporting.

Only Patton was in the helicopter when the ambush occurred, the point of greatest criticism at the later enquiry because despite the emotion of the discovery Bartel shouldn't have left the machine, any more than Colonel Forest should have ignored the elementary precaution of posting pickets, instead leading his men into a hurried evacuation line. Everyone formed part of the line, even his father and Vine; only Lind remained uninvolved, recording it on film. And became the first casualty, hit by the initial burst of fire from the North Vietnamese unit which had encircled them. Blair had been nearest the helicopter, carrying a child, able to run the last few yards, lay the baby with those already rescued and get to the minigun and beat back the attack threatening to engulf them.

His father and Peterson made it too, shielding the orphans with their own bodies. But able, from what his father wrote, to see those following with the remainder of the babies cut down in the concerted crossfire. Patton, who risked his life hauling the dying Lind into the helicopter, made two runs to ensure none of their group survived before scurrying at tree-top level back to Saigon.

The award to his father of the title of International Reporter of the Year was for the consistency of his coverage but the file on the orphan rescue was the individual story quoted in the citation. And cited again in that special commendation by the Pulitzer committee, a unique tribute to an Englishman from an exclusively American press award organisation.

Deservedly so, Hawkins decided. At this stage in the planning of the book it was impossible to make positive decisions but if the construction allowed he intended running in full his father's account of Chau Phu. Maybe even start the book with it: certainly include it near the beginning. His father's career had begun in Asia and ended – actively – there, so it would be fitting to do so.

Hawkins squatted in the basement, gazing strained-eyed around at the albums he had to assemble into an orderly narrative, together with the correspondence – that of his father to him, not just the letters to and from the famous – and the memorabilia. He had enough for a frame-work, to satisfy Wilsher and Jones and presumably Doondale. But frame-works were precisely that, outlines upon which complete pictures or shapes were created. He needed more, much more, if his book were to be properly recognised for what he wanted it to be. Early though it was, he already knew the sentence with which it was to begin, however. He pulled the first of the index cards towards him, deciding it was the appropriate way to begin the notes.

My father was a hero, a man of whom I'm proud, he wrote.

There were three beers left unopened when he rose to go upstairs. I'm proud of myself, too, he thought.

Eric Patton had come to look for omens in the relationship and had hoped the memorial dedication might provide one; that the visual evidence of her husband's death carved in stone might enable Sharon Bartel to accept the finality of it. He'd never tried to force the pace and he didn't now. He brought his own aircraft from New York and made no attempt to intrude into her silence, either on the return flight or during the ride up into Connecticut. He didn't suggest he should stay overnight, either, driving back into Manhattan despite it being late.

He went up again at the weekend, as he always did, and spent the Saturday doing what he enjoyed most, looking after her. He unjammed the blocked waste disposal and checked that the snow tyres were in good shape and that there was a booking for her at the garage to get them fitted, before the winter fall which was soon due.

It was Sharon's suggestion that they eat in and Patton was glad because he wanted the opportunity. While she cooked he stacked the fire and opened wine and waited until the dessert, because he wanted the timing to be right.

'Well?' he said at last.

'Don't, Eric.'

'Why not?'

'You know why not.'

'He's dead, Sharon. He's been dead for seven years.'

'I know.'

'Then why not?'

'Because,' she said, child-like in her misery.

'Don't you love me?'

'We've talked about this a hundred times: you know I love you.'

'And you know I love you. We sleep together and we enjoy being together and we *are* together. It's ridiculous for us not to get married.'

'I know,' she said, in weary resignation.

'Why then?' he repeated.

'It just seems . . .' Sharon shrugged her shoulders, seeking a familiar escape. 'I can't avoid thinking of it as a betrayal.'

'You can't betray a dead man, darling,' he said.

She winced, gazing down at her discarded plate. 'No,' she conceded.

'So marry me.'

'Give me some time,' she pleaded.
'I've given you all the time in the world!' he said. 'Years!'
She smiled up at him at last. 'And I love you for it,' she said.
'So?'
'Soon,' she promised. 'We'll decide very soon.'

Chapter Four

The Senate office buildings were on his side of Capitol Hill so Hawkins decided to walk to his meeting with Peterson, despite the winter cold. The wind had long ago stripped the rattling trees and he bent against it, almost at once feeling his eyes prick with tears. He scrubbed against them, wishing after a few yards that he had taken a cab, despite it being such a short distance. He actually looked both ways along the avenue, seeking one, and then shrugged, continuing on: at least the walk gave him time to think through the meeting with the senator.

Will Peterson Declare had become the most popular political guessing game in Washington. And not just Washington. In the time it took to fix the meeting suggested at the memorial dedication, fresh speculation had appeared right across the country, in *US News and World Report, Newsweek, Time* and the *Washington Post:* the *New York Times* had actually carried a Gallup poll commissioned to estimate his popularity over every other possible candidate within the Party. Peterson had come out with a thirty percent lead over his nearest rival.

Which made the comparison feature between Peterson and the incumbent Harriman very apposite. And therefore professionally necessary to retain London's confidence that he could do both things. Hawkins had become utterly absorbed in the structuring of the book about his father, and managed to get the outline to Jones before the man demanded it. There'd been praise – which he guessed was begrudging – and a request for sample chapters, three of which were almost ready. Hawkins decided to tell Peterson; seek his reminiscence and help, but only at the end of the meeting when he'd got as much as he could about the man's presidential ambitions. From the press building gossip Hawkins knew he was the only British journalist so far to have gained access, ahead of at least five other London interview requests. It was the sort of thing his father would have achieved.

Hawkins went gratefully into Dirkson building, pulling the hunched collar of his coat down and submitting his briefcase to the familiar security check. There was an appointment confirmation at the desk and as he approached the suite towards which he was directed Joe Rampallie emerged, blinking expectantly along the corridor.

'Glad you're on time,' said Rampallie. 'Senator's got a very tight schedule.'

Rampallie's eyes shuttered behind his impressive glasses and Hawkins wondered what job the man would be awarded under the patronage system if Peterson made the White House. He hoped it was one the man's nerves could handle.

'How much time have I got?' he said.

'Thirty minutes,' said Rampallie. He raised his hand in a measuring gesture, and said, 'We've got press applications up to here.'

The outer rooms of the suite were disappointing, like the Vietnam monument had been in such grandiose surroundings, cubicles partitioned by plaster-board with just sufficient room for a person, chair, desk and telephone. Only Peterson's office matched the expectation. It was a panelled, high-ceilinged room, with windows overlooking the Dome. Directly in front of the windows was an impressively large, inlaid desk, with the American flag furled in its pod alongside. Against one wall were glass-fronted bookcases and above them a gallery of pictures, featuring Peterson through his college days to his entry into politics: Hawkins saw at least four showed the man during the Vietnam visit. Used and comfortable leather settees and chairs were arranged against the facing wall and in front of them, seemingly haphazard, were set several tables: magazines and newspapers were disarrayed upon two, as if someone had been hurriedly going through them. It had the appearance of a busy place.

Peterson was behind his desk, apparently working, when Hawkins entered. The politician rose immediately and strode across the room to meet him, hand outstretched. 'You look cold.'

'I walked. It was a mistake,' admitted Hawkins.

'Coffee then?' He paused and added, 'Or something stronger?'

'Coffee would be fine,' said Hawkins, who had discerned the hesitation.

Peter Elliston had followed Hawkins into the room from the outer offices. Peterson nodded towards the man and said 'Me too.' Then he indicated the couches and as they sat said, 'Pity we haven't seen each other for so long.'

'It's been a busy time for you,' said Hawkins.

Peterson smiled and said 'Maybe getting busier.'

'I want to do a comparison piece, between you and the President,' announced Hawkins, seeing the opening.

'It's been done before,' said Peterson.

'Not in Britain; and not in the depth I want to do it.'

'That sounds like your father,' remembered Peterson.

'I want to talk about him, too,' said Hawkins. Keeping to the priorities he added, 'But later.'

'Why don't we give it a try?' agreed the politician.

Hawkins took the tape recorder from his briefcase and said to Peterson, 'Do you mind these?'

'Never object to an accurate record,' said Peterson.

Hawkins pressed the 'start' button and said, 'I've read all the cuttings and the hand-outs but I'd prefer to get it directly from you: an accurate record.'

Peterson handed him a coffee cup, smiling again at having his remarks returned to him. He settled back against the leather upholstery and said, 'I'm forty-four years old. Forty-five in June. Father was Walter Peterson: mother's name was Jane. He was chief accountant with a lumber company in Charleston . . .' The man raised his hand, warningly. 'That's Charleston, West Virginia, not South Carolina. Big house, couple of servants. I was in my last year at Yale, already with a place at Harvard when the investigation started into the company accounts. There was a discrepancy of $300,000. My mother found my father hanging in an outhouse. I quit school and went back to Charleston. Sold the house and was able to pay the company back $50,000 right away. I kept back $50,000, after all the usual debts were settled and used that as a stake. I'd been doing engineering at Yale and so I decided upon electronics, which was new then: an unknown quantity. I was lucky, getting in so early. Never did go along with the idea that all the Japanese could do was copy and badly at that. Amalgamated with an Osaka firm after four years, which gave them the benefit of an American outlet and me the advantage of their research facilities. Which meant we got early into microchips. Paid the full debt back to the lumber company within five years of setting up the operation. Got interested in local politics and then decided to try it nationally, just once. If I'd failed I'd have stayed where I was. But I got in, first time. Lucky, I guess. Met Eleanor here in Washington: her father was Harry Black, leader of the House for eight years. Been married for ten. Amelia is eight, John is six . . .' Peterson stopped, drinking heavily from his cup. 'That's about it,' he said.

It wasn't thought Hawkins, but it was a beginning. He said, 'There was never any prosecution against your father?'

Peterson shook his head. 'He committed suicide, like I said.'

'So there was no need to make it public. I would think a lot of people, particularly a politician, would have wanted to keep it quiet.'

'And regretted it, particularly a politician,' said Peterson, coming back at the other man. 'You know what politics are like, particularly at this level. Sure I could have covered it, made up a story why I quit school. And then somebody would have started to dig and there would have been an exposure. I think I would have suffered far more politically by doing it that way than by facing the facts as they exist. And as they would have been discovered to exist.'

'A lot of cynics would say such openness doesn't fit into politics.'

'I don't think much of cynicism but I take your point,' said Peterson. 'I'm not offering myself as some shining white virgin. I've got it wrong, a lot of times. But never by cheating. Or by avoiding unpleasantness, hoping it would go away.'

'Your Vietnam stance was unpleasant.'

'My Vietnam stance was *right*!' insisted Peterson, vehemently. 'Let's consider all the facts about Vietnam and let's look at them objectively. America was in Vietnam long before the Tonkin resolution in 1964. It was the longest conflict in

American history. Nearly 2,700,000 Americans fought there: 300,000 were wounded and 75,000 permanently disabled in addition to the 58,000 killed that we commemorated the other day. And it wasn't anything to do with them. They were serving their country, doing what successive administrations . . . administrations of both parties, so I'm not making political points here . . . ordered them to do. And you know what happened? When they came back wearing their uniforms and their medals, thinking they'd served a grateful country, they'd be stopped in the streets and asked how many innocent civilians they'd murdered. People would actually spit at them: the advice went back down the line 'Don't come out wearing your uniforms: they'll shit on you . . .'

'Harriman suffers from the accusation that he's been outmanoeuvred by the Soviets: that he's weak, in confronting them?' said Hawkins.

'That's his misfortune,' replied Peterson, achieving just the correct amount of political qualification.

'How does your attitude towards what happened in Vietnam reconcile with your apparent belief that Moscow has to be confronted by strength?'

Peterson frowned, as if he had difficulty with the question. 'The word "reconcile" indicates the bringing together of diverging points and I don't think my attitude toward the Soviet Union is in any way in conflict with my record on Vietnam. In 1954 this country had the evidence of what had happened to the French to prove that Vietnam was a nationalistic war. It was as obvious as the nose on your face. But it was the time of paranoia about communism . . .' He put his hand up, in another warning gesture. '. . . remember the word, paranoia. It was the time of Dulles and the domino theory – if one country goes, the others will fall with it. Our involvement made it what it wasn't, an East versus West, Moscow versus Washington situation.'

'Does that mean complete non-interference?' pressed Hawkins.

'It means properly regarding international developments for what they are, not what some hobby horse of the moment dictates what they might be.'

'Does that mean you support American involvement throughout Latin America?'

'Not by a million miles,' said Peterson. 'I don't think this country has done itself any good at all, either in Latin America or in the uncommitted Third World, in backing some of the Right Wing regimes that it has. It's another example of paranoia, imagining one extreme and going to the other. Do you know America spent in excess of $20,000,000 preventing Salvadore Allende retaining power in Chile and a later enquiry headed by Vice President Nelson Rockefeller found, among other things, that after it was all over there never was any significant danger of Allende-style communism infecting the rest of Latin America!'

Rampallie appeared at the door of the office and Peterson looked enquiringly at him.

'We're not through,' he said to his campaign manager.

'It's a tight schedule,' said Rampallie.

'Loosen it a little,' ordered Peterson.

Hawkins watched the disgruntled man retreat from the office and then said to Peterson, 'What about America?'

'Stalled by the wrong policies,' said Peterson. 'We've had monetarism long enough. It's shaken out the dead wood, the industries and the corporations that didn't deserve to survive anyway. Now we want a reduction of interest rates and some government-sponsored expansion schemes. I'm not making any New Deal comparisons but unemployed on the streets of America isn't a sight that pleases me: not one that pleases me at all. We're fast approaching a chicken and egg situation where welfare benefits and unemployment pay is costing more than the inflation the present administration is so committed to reduce. That isn't even logical, let alone sensible.'

Hawkins paused, considering the approach. Then he said, 'I think I've just heard the views of a presidential candidate.'

Peterson's grin came again, a lopsided expression that with the flop of hair gave him a strangely boyish look. 'When's this piece for?'

'Sunday,' said Hawkins.

'Guaranteed?'

'Nothing's guaranteed in a newspaper, not until publication.'

Peterson nodded, apparently considering something. Then he said, 'I'm in California on Saturday: $100 a plate fund raising dinner in Los Angeles. I'm making the official declaration then. The Party chairman and most of the committee will be there as well, conveying the impression of endorsement before the convention. The campaign starts on Monday.'

Hawkins felt a numbed moment of excitement. 'Nothing official until then?' he said, wanting an absolute assurance.

'Nothing,' said Peterson.

With the eight-hour time difference between Los Angeles and London, Hawkins knew it gave him a world exclusive that no other Sunday newspaper could match, providing they put a copyright slug on it. 'Thank you,' he said inadequately. 'Thank you very much indeed.'

Rampallie made another impatient appearance at the doorway. Peterson splayed his hand out towards the man, indicating a further five minutes and said, 'You spoke earlier about wanting to talk about your father.'

Quickly Hawkins explained the idea of the biography and of his intention to begin the book with a chapter on the orphan mission.

'I'm not sure I can help you beyond what you already know,' said Peterson. 'It was fully reported. And the enquiry was made public.'

'You were there,' said Hawkins. 'I'd like to hear your personal recollections.'

'Why not call me at Georgetown after the weekend and we'll talk about it some more?' said Peterson.

Hawkins rose, finally. At the door he turned back to the politician and said curiously, 'Why me? Why a British paper? There are a lot of influential commentators here in America from whom you'd get greater advantage.'

'I liked your father,' said Peterson simply.

How long would it be, wondered Hawkins, before he got things for himself?

The flight from Fort Bragg came into the military airfield at Andrews. There was a helicopter waiting and Blair went straight to it, buckling himself in with accustomed expertise. As it lifted into the air and skittered out over the bare Virginia countryside he thought, as he often did, how different it was travelling in an enclosed machine compared to the open-side gunships he'd become so used to in 'Nam. Gazing down, he thought idly that there was a passing resemblance to the frozen winter landscape and some of the defoliation he'd known there, too.

Soon they picked up the familiar thread of the Potomac and Blair began isolating landmarks, the Washington Monument like some giant compass point in the middle of the city, then the Capitol and the Lincoln and Jefferson commemorations. The Colonel turned away, looking more directly downwards over the suburbs of Rosslyn and at the familiar five-sided shape that gave the Pentagon its name. America's military headquarters looked exactly what it was supposed to be, he decided: a block-house from which to command wars. It was unfortunate there hadn't been any, for so long. Too long.

He nodded his thanks to the pilot and followed the waiting orderly through the interlocking corridors, grateful that he had a guide despite the numerous occasions he'd been there.

Griffiths was waiting in the middle of his office when Blair entered, smiling expectantly. The General had been his senior by two years at West Point, where the friendship had begun there and endured, despite long periods – sometimes years – when they'd been in different parts of the world.

'Good to see you, Elliott. Good to see you.'

'And you, Greg.'

'Glad you could come up,' said Griffiths.

He was a white, grizzled-haired man who had just started to let himself go, with a beginning of a sagging belly tight against his uniform. Blair, whose weight at forty-six was the same as it had been when he was twenty, thought it was unfortunate. He never permitted any relaxation in himself. 'You said it was important,' he reminded the other man.

Griffiths led his friend to a chair, prolonging the announcement. 'I wanted to be the one to tell you, unofficially of course,' he said. 'You're going to get it, Elliott.'

Blair allowed a brief smile. 'You sure?'

'Put it in the channels myself. There's the usual red tape but there isn't a chance of it not going through: I've never seen more unanimous recommendations. You've got a lot of friends where it's important to have them. What's it feel like to be a General?'

Blair didn't answer at once, thinking. Then he said, always controlled, 'Good. Very good indeed.'

Chapter Five

The Sunday edition led with the exclusive of Peterson's declaration and the accompanying profile ran across two inside pages; American newspapers took the announcement from the senator's speech in California but there was wide-spread syndication of Hawkins' feature. There were several cables of congratulation from London, including one from Doondale. Wilsher's herogram included the line 'like father like son'. There was nothing from Harry Jones.

Determined upon the separation of activities, Hawkins devoted full attention to the biography in his spare time. He wrote to Elliott Blair at Fort Bragg requesting a meeting and on the first available day arranged the schedule carefully for his other meetings. Eric Patton seemed more reserved during their telephone conversation than he had at the dedication ceremony but finally agreed to an afternoon meeting at his Manhattan office. It was the timing for which Hawkins hoped because it left the morning free for the CBS film, for which they agreed to make a viewing room available by eleven o'clock. Deciding fully to fill the free day, he finally called the Georgetown number. It was Eleanor Peterson who answered. There was a muffled conversation from her end, away from the receiver and then she came back to say an evening meeting would be fine.

Hawkins replaced his telephone and sat back content. If he could see Harvey Lind's film and speak to Patton and Peterson all in one day that only left Blair. He'd airfreighted the sample chapters Jones had demanded and the indexing was fairly well advanced, with just the cross referencing of the personal correspondence to be completed: soon, Hawkins thought, he would be able to start writing properly. He was anticipating the moment.

Hawkins caught the eight o'clock shuttle from National to allow for any delays but there weren't any and he was in Manhattan before ten. He thanked the CBS editor, Art Shaw, for making the viewing possible and the archivist, whose name he didn't quite catch but thought was Mort, for going through the records.

'No problem,' assured the man. 'It's a pretty well known film.'

'I've never managed to see it before.'

The archivist frowned. 'Thought we made a copy available, for your father.'

Hawkins shook his head. 'I've got everything in Washington; there's no film.'

Shaw showed him into the tiny viewing room and explained how the button set into the arm of the chair would automatically start the video already waiting on the machine. It was the complete, uncut film, actually beginning with John Vine giving editing instructions into camera and the date as 24 April, 1975. The CBS reporter was a fat man who appeared bothered by the heat. Before beginning his commentary he wiped his face with a handkerchief and straightened the crumpled safari suit. Hawkins started looking for his father, beyond the television reporter. Vine was standing in front of a helicopter against which Hawkins was able to identify Patton, much slimmer than he'd been in Washington, carrying a helmet and wearing black, wrap-around sunglasses. Another man also carrying a flying helmet and whom Hawkins presumed to be Charles Bartel was talking to a group of soldiers, all in camouflaged jungle greens. The Green Beret support group, Hawkins supposed: as he watched a tall man wearing Colonel's insignia moved slightly, confirming Hawkins' impression by revealing Blair. The man on the film appeared unchanged from the person whom Hawkins had met, seven years later, at the dedication ceremony. He still held himself stiffly upright, still slightly apart from whatever conversation was being conducted. From the insignia Hawkins knew the colonel to be Forest but wondered which of the other two soldiers was Marne and which was Lewis. And where, Hawkins wondered, were his father and Peterson?

'This is Vietnam, in the final agonising days of a long and bloody war,' said Vine, beginning a report from which he never signed off. 'And this is Tan Son Nhut, the vast airport complex on the outskirts of Saigon which has for so long been one of the centrepieces of that war. But today, from Tan Son Nhut, there is going a mission not of war but of peace . . . a mission for which one man – to many an unpopular man – the controversial Senator John Peterson, is responsible . . .'

Peterson came into shot, from what Hawkins realised was a pre-arranged cue. As the politician approached the helicopter, Vine stepped up and said 'Can you tell me, Senator Peterson, what you are doing today?'

Peterson turned more fully into the camera, squinting slightly against the sun. Lean then, thought Hawkins, and still with that uncontrollable shank of hair.

'It's no secret to anyone in America why I'm here,' said Peterson. 'I'm here because I think that although America is not wrong in getting out of this misbegotten, unwinnable war, it is wrong in abandoning so flagrantly its allies, just as it has, regrettably, abandoned back home on the streets of America loyal citizens who have come to fight this war. There is little I can do to prevent that abandonment or that attitude, appalling though I consider both to be. But there is something I can do to avoid it being quite so complete an abandonment. I have come to Vietnam not just on a fact-finding mission upon which I intend to report fully from the floor of the Senate. I have come with locations of orphanages maintained by American charities from which there are grave fears that local support staff might have fled. It is my intention to try to locate some of those orphanages and spare at least those children, most of whom are of American parentage, falling under communist domination.'

Hawkins wondered how many times the reporter and the politician had rehearsed the statement.

'Where, exactly, is the orphanage you are attempting to reach today?' asked Vine.

'Five kilometres outside a place called Chau Phu,' said Peterson. 'Quite near the Cambodian border.'

'Senator, I've had it suggested to me that the Vietcong actually have positions around the perimeter of this airfield: won't Chau Phu be in Communist hands?'

'Intelligence reports are that the place is quiet,' said Peterson.

'Thank you, Senator,' said Vine. He visibly relaxed, turning to the camera and saying to the unseen Lind, 'OK?'

There must have been some affirmative gesture from the cameraman because Vine nodded to Peterson, who continued towards the helicopter.

And then Hawkins saw his father. Hawkins realised that professionally the man must have kept out of shot until the opening was completed. The stance and the demeanour were different from what Hawkins remembered: Edward Hawkins walked with a controlled authority and as he watched, the son saw that it was to his father that the waiting crew and soldiers appeared to turn, rather than to the senator. Lind began to take the shot in tightly, zooming upon the actual departure preparations as Bartel and Patton got into the machine, followed by the Green Berets. There was a nodded exchange between Peterson and his father and briefly the two men remained outside the helicopter.

Hawkins came slightly forward in his seat to study the film before him. His father was only an inch or two shorter than Peterson although heavier, the middle-aged paunch obvious despite the tailoring of the shirt. But although it was receding there was no grey in the sandy hair. Hawkins wondered if there was the similarity in looks and build between himself and his father which people so frequently remarked; he could never see it.

Peterson, then his father and finally Vine were shown entering the helicopter and then abruptly the film cut, for what Hawkins presumed to be Harvey Lind's entry into the machine. It began again with equal abruptness, the camera focussed over the shoulder of an unidentified gunner, the minigun almost filling the frame. Gradually Lind projected his lens beyond, to cover the ground below.

Hawkins had seen film of Vietnam before but never ceased to be caught by its beauty; and this was particularly good film. It was still very early in the morning, which Hawkins knew from his father was always departure time. Thick white mist was puddled like milk in the dips and valleys. And not just dips and valleys, Hawkins saw: pock-like bomb craters, as well. The jungle ceiling was tight packed, so closely that Hawkins was reminded of looking down at a cauliflower. Hundreds of men could have moved undetected beneath that sort of cover he thought. And had done so. Occasionally, like leprous scars, there were burned, bare patches where the defoliant had fallen, destroying everything except the grey stumps of dead trees which stuck upwards like lunar gravestones.

There was another short moment of blackness and then the location shots began, as the helicopter came in over Chau Phu. The helicopter made several runs over a straggle of thatched huts and then a paved road leading into the town proper. Lind changed angles, bringing the camera back inside the helicopter to show Peterson staring down intently: just beyond the American, Hawkins saw his father again. He was writing something in a notebook.

Lind switched back to the countryside below and Hawkins thought of Peterson's filmed assurance; everything certainly looked pastorally quiet in the outskirt hamlet.

Lind kept the camera running throughout the descent, even when the downthrust of air showered elephant grass debris and dirt all around them, making it difficult to see. The Green Beret group leapt out into the dirt fog and Lind went with them, the focus jerking up and down as he ran. There was another break and when the film began again the dust had settled. The soldiers had already probed a perimeter around the machine and were walking back towards it when Peterson, Vine and his father emerged. In single file they tracked in towards the Vietnamese huts, and what now, from ground level, could be seen to be a brick-built structure.

The film cut directly inside and Hawkins blinked at what he saw. What was described as an orphanage was a shell of a building, with no division of rooms. There was no furniture of any kind, just matting upon the floor and upon it, packed tight, a jumble of babies. Lind was shooting with the sound running and Hawkins heard Peterson exclaim 'Jesus Christ!'

And then he became aware of something else. There was no sound from a room full of abandoned babies: they had been left so long and were so near death that there was no crying left among them.

The film showed a flurry of movement and disjointed sound, Peterson at the doorway gesturing to the helicopter for help and shouting 'Come here' and three figures emerging from it.

Then a scurrying line of men, cradling the stick-limbed children from the building and into the helicopter. Lind moved with the line, right up to the helicopter, where Patton crouched trying to assemble them in some kind of accommodation on the hard metal floor. Lind changed shots and Hawkins guessed the cameraman was against one of the skids now, shooting back towards the rescuers.

It was a fast moving line, because the children had no burdening weight for the men: a gentle run from the orphanage, carefully laying the baby into Patton's waiting hands, and then a faster burst back to collect another tiny body. There was no unnecessary conversation, just the growing panting of running men. Hawkins recognised the voice of the unseen Patton saying 'There! There!' to babies who could not understand and were beyond hearing anyway.

Hawkins heard another voice he didn't recognise say 'got nearly all of them now.' The shot was of the line of men coming towards the machine, Blair leading, then Peterson. Two Green Berets had their backs to the camera, returning to the orphanage. Another was third in the line of the approaching men and then came a brightly-shirted Edward Hawkins. The outbreak of firing was hard to detect, indication coming not from any sound but by a sudden almost imperceptible halting of the line. Blair, the trained professional soldier, started forward at once and another unidentified voice yelled 'Incoming; ambush' and then the film careered wildly out of control, a glimpse of the helicopter and then blackness, as Lind was hit.

Hawkins needlessly pressed the button in his chair-arm, to stop the cassette and then remained slump-shouldered in his seat, surprised at his own emotion. His eyes were wet and embarrassed at the possibility of being discovered crying by Shaw or anyone else he quickly wiped them and then blew his nose. He recovered before the door opened behind him: both Shaw and the archivist came into the viewing room.

'Some film?' said the archivist.

'Tremendous,' agreed Hawkins.

'Deserved the award,' said Shaw. 'Poor bastards.'

'I checked while you were watching,' the archivist said to Hawkins. 'According to our records we did make a copy for your father. Went to him in England around November, 1975.'

'He left a tremendous amount of material,' said Hawkins. 'I must have missed it: I'm sorry.'

'No trouble,' assured the archivist.

'Like to see a copy of the book when it's published,' said the editor.

'I'll make sure you get a copy,' said Hawkins. 'Both of you.'

Hawkins emerged thoughtfully from the CBS building, halting directly outside and standing back against the flow of pedestrians. He'd been through every one of the trunks in which his father's records were stored: indexed it too. He knew there wasn't a video cassette of the Chau Phu ambush.

The name plate said simply 'Eric Patton and Co.' with no indication of the business so it was not until Hawkins emerged on to the eighteenth floor of the Third Avenue skyscraper that he recognised from the wall to wall mountings that it was a freight company.

But no ordinary freight company. The expansive vestibule was lined with photographs of aircraft, helicopters and trucks, all emblazoned with Patton's name and the area behind the receptionist carried illustrations of ships which Hawkins presumed to be also part of the empire.

The receptionist was expecting him and within minutes an austere, bespectacled woman appeared from a side door to escort him. Beyond the boxed-in entrance area Hawkins saw the work section was arranged open-plan and occupied the entire floor: from the elevator listings, Hawkins remembered Patton occupied the storey above, as well. At least fifty people were working at serried rows of desks: against a far wall were banks of telex machines and above them clocks, with capitals named above, recording the time differences around the globe. As he passed Hawkins saw that every major trading nation in the world was recorded.

Patton's office was separated off from the main working area, with two outer secretarial suites dividing it. It was the most impressive of all. It occupied a corner position, with a view in two directions over Manhattan; Wall Street and downtown and the twin Trade Towers to the right, the United Nations building East River and in the distance Brooklyn behind where the man sat.

The desk matched the office in size. At its top was a boxed

line of trays into which documents and papers were neatly partitioned and at either end were miniature flags: one was blue and white and had the initial 'P' at its centre and the other was the American flag. There was a telephone console on a special table arranged at right angles to the main desk and a single photograph frame, against it. There was a small conference table and chairs to the right of the door through which Hawkins entered and couches and easy chairs in front of the facing wall. Like the vestibule, those walls were occupied by more pictures of ships, aircraft and trucks.

Patton emerged from behind the desk with the nervously hopeful smile Hawkins remembered from their first meeting. The woman remained in the room and Patton looked to her, then to Hawkins. 'Can I get you anything?' he offered. 'Tea? Coffee? Drink?'

Only beer in the basement, nothing last night except wine with a badly-cooked steak and so far nothing today, Hawkins counted off triumphantly. He could afford to relax. 'Scotch,' he said. 'Thank you.'

'Me too,' said Patton.

The woman went to an inset row of cupboards and Hawkins saw there was no measure on the bottles. She poured generously into both glasses and delivered them with separate, diminutive water jugs and accompanying paper napkins. As she served Patton said, 'What can I do for you?'

Hawkins went into what was becoming a well-rehearsed explanation about the book, aware as he spoke that as well as the smile Patton had the mannerism of nodding his head in either attentiveness or understanding to what was being said to him.

When Hawkins finished the American said, 'I don't really see how I can be of any help. I never met your father until that day. There was a lot of publicity immediately we got back: went on for a day or two, with the kids, but we didn't really get together. There was the enquiry, but our evidence was given on separate days, so we didn't really see much of each other.'

'What about the ambush?'

Patton turned down the corners of his mouth, a doubtful expression. 'Been gone over a hundred times,' said the man. 'Let's admit it, we were careless. Once we found those kids in the state they were in, the only thought was to get them out.

The 'Cong were all around us and we didn't even know it.'

'I've just come from seeing Lind's film,' said Hawkins. 'Tell me what happened afterwards.'

Patton wasn't smiling now. 'A lot of confusion, trying to understand what was happening. Lind was hit bad, in the head and in the chest, too. I pulled him into the helo by the straps of his camera harness. He was still breathing but obviously in a bad way. Blair knocked me over, coming in. He was carrying a kid. He put it down with the rest and got immediately on to one of the guns . . .' The man paused, shaking his head in admiration. 'Jesus, if ever a man deserved the citations and the medals it was Blair. The main assault seemed to be coming from the port side: by the time he got to the minigun we could actually see them, not more than twenty yards away. He just cut them down: wiped out that first attack. If it hadn't been for him, they'd have got us all. He was going back and forth across the machine, stepping between the kids, blasting away first from port, then to starboard, then back to port again. God knows how many he killed. Dozens I guess.'

'What about Peterson and my father?'

'Peterson got in first: he was in the line behind Blair. Then your father.' There was another hesitation. Patton continued, 'Blair took command: we were both majors, but he had seniority. He yelled at Peterson and your father to get down and out of the way. The only way they could do that was lay over the kids, to protect them. So they did. They didn't fight, like Blair was fighting, but they were just as brave.'

'Why did Blair go back and forth?' wondered Hawkins, curiously. 'Why didn't he man one gun and you the other?'

'I was trying to help the rest of the line,' said Patton. 'I had an M-16. I was trying to cover them to the front, so they could get through.'

'Why didn't they?'

Patton frowned at the directness of the question. 'It was the consensus of the enquiry that the 'Cong realised what we were trying to do: they were pouring fire in at that stage not so much *at* us but between us and the others. Creating kind of a screen, I guess.'

'The others had no protection?' said Hawkins.

Patton shook his head, head bent in recollection. 'Colonel Forest had them organised: always felt it was a damned shame that he got the censure he did. Like I said, the main attack was coming from port, from the left. He and his men formed up against that, with my guys – Bartel and Chaffeskie and McCloud – covering their backs. The kids they'd been carrying and the television reporter were in the middle of the two groups . . .'

Hawkins saw the man was squeezing one hand over the other, white-knuckled, under the emotion of the memory and there was a shoulder movement which made him wonder if the American were crying.

'. . . They were trying to make it crabwise to the helicopter,' took up Patton. 'But there were the kids: they couldn't move. The reporter, Vine, had an armful of them: a goddamned armful . . .' Patton extended both arms in an embracing gesture. '. . . Like that, like he was gathering in corn or a bunch of sticks or something. And he was staggering between these two lines . . .' He stopped, shaking his head. 'Christ, it was awful . . .'

Hawkins realised that the other man *was* weeping. And that he should stop the questioning. 'What happened?' he went on, instead.

Patton breathed deeply, trying to control himself. 'It was Vine who got it first,' he said. 'He was standing more upright than the rest: he had to, to be able to carry the kids. Right in the head. I saw him fall and all the kids with him. Just like pins in some goddamned bowling alley. They fell into Bartel and the gunners, breaking their fire. Bartel got it next: like I said there was a hell of a lot of confusion. But I'm sure it was Bartel. Then I think one of the Berets got it: not Forest, one of the others. It seemed very quick . . .' He shook his head. '. . . I don't remember who got it after that. Too quick . . . so very quick . . .'

'What then?'

'I shouted that they were down. You know about those miniguns? Hell of a rate of fire. Blair had used up the ammunition on the starboard gun and was getting low on the other. The 'Cong were closing in again. We could actually see the bastards. You know they really did wear those black pyjamas, like all the stories said! We knew we couldn't get to our guys, not on foot. So I took the helicopter up. When they saw the rotors start the bastards became suicidal, just running at us and disregarding the gun, to stop us. Blair managed to reload, just one side – the port I think – I flew sideways into them and he was firing into them, blowing them away like targets on a funfair. I fucked them, doing that: went right over their goddamned heads. Gained height that way and then came back, so we could go directly over our guys . . .'

Patton hurried a handkerchief from his pocket and blew his nose. '. . . They were gone,' he said. 'I made four passes; went as low as I could. There wasn't a movement. Nothing. They'd been cut to pieces. We'd taken a lot of fire: lucky as hell not to have been hit. The engine was missing and I didn't know if it had taken a shell or something. Once we knew they were dead we came out. Wanted to get some height, to get out of range, but there wasn't sufficient lift. Learned later that the rudder bar had taken a hit and was distorted. Came back to Saigon just above the trees and shitting myself the whole way that someone would open up from below and bring us down. Nothing happened, though. After we got back to Saigon someone counted eighty-five hits against the helicopter . . .' Patton looked up, swallowing. 'That's it,' he said. 'That's how it happened. Just like your father said.'

Hawkins, who by now knew his father's account very well, decided that it wasn't just like his father had said. This was much more dramatic. Perhaps, for once, his father had been overly objective. 'I'm sorry,' he said to the other man. 'I didn't know it was going to be like this.'

Patton's ever-ready smile came on, embarrassed this time. 'I think I need another drink,' he said, standing. 'You?'

'Yes,' accepted Hawkins, standing with him.

'Scotch again?'

'Fine.' Still only the second: no problem.

Patton went to the low bureau against the wall beyond the conference area. Hawkins wandered in the direction of the desk, staring out over the haze-smudged, broken-toothed skyline of Manhattan towards Brooklyn. He'd definitely make the Chau Phu episode the opening of the book, he decided. 'That was a hell of an experience,' he said.

'It was,' said Patton, from behind.

Hawkins turned back into the room, to face the man again. Patton was on the far side, still in the conference area and the large desk was between him and where Hawkins stood, near the panoramic window. Instinctively Hawkins' eyes fell to the photograph in the single frame and he recognised Sharon Bartel at once.

Patton saw the look and came angrily across the room.

'Get to hell out of there!' he said.

'I'm sorry,' said Hawkins, embarrassed. 'I didn't mean . . .'

Patton slammed the drinks down against the desk, all friendliness gone. 'I don't want her in any damned book. I don't want you prying into my life or hers: exposing her to any more pain than she's already suffered. Try some half-assed trick like that and I'll sue you for whatever my lawyers can come up with and block publication in any country you try to get it printed. Any country, you hear me!'

Hawkins physically moved back against the window, feeling it cold against his back, bewildered by the outburst. 'Any friendship between you and Mrs Bartel isn't of interest in what I intend to write,' he said anxiously. 'It's about my father, nothing else. You're part of it because of my father, not because of her. I'm not intruding, into anything. Believe me I'm not.'

Patton relaxed slightly. 'You're not shitting me?'

'No,' said Hawkins. 'I'm not shitting you.'

'Your word?'

'My word,' promised Hawkins. 'And if that isn't good enough, you can read the finished manuscript, before it even gets to the publisher.'

The smile returned. Patton offered the Englishman the delayed whisky and said, 'I'm sorry: guess I ran off at the mouth a little.'

'I understand,' said Hawkins. Maybe a newspaper story, he thought. But not for his newspaper. And anyway today he was working on the book, nothing else.

'I just don't want her hurt any more. No more pain.'

'I said I understand,' repeated Hawkins. 'And I meant the undertaking about the manuscript.'

'Your word is enough,' said the American. He drank heavily from his glass. 'Vietnam was an asshole,' he said, in bitter reflection.

'Yes,' said Hawkins, recognising the inadequacy of never having reported any war. 'I suppose it was.'

'Sorry I blew up.'

Hawkins managed to get out of Manhattan ahead of the evening rush hour and make the shuttle connection he wanted at La Guardia. He sat back in his seat, eyes closed, for the return flight to Washington, drained by the day and aching, physically tired but not drunk. How long, he wondered, before he would be able to stop counting and know he'd conquered it?

He took a cab directly from the airport to Peterson's house. Georgetown is a favoured political suburb of Washington, the threads of Constitution Avenue and M Street knitting together a backstreet lacework of exclusive town houses. Peterson's home was a four-storey brick and clapboard building on Dumbarton Street, fronting right against the pavement. Hawkins sounded the bell and having been there before and knowing the system looked directly up into the lens of the television security monitor. After several moments Eleanor Peterson opened the door, looking at him curiously.

'John's not here,' she said. 'He left a message on your answering service.'

'I've just got back from New York: I haven't called my service,' said Hawkins. He felt foolish, making the admission to the woman.

She appeared uncertain and then said, 'Why not come in for a drink anyway?'

Hawkins hesitated in return. Only two so far, he remembered. 'Thank you,' he accepted. He thought she looked beautiful: but then he'd thought that for a long time. And knew he shouldn't.

Patton remained unhappy at the Englishman's identification of Sharon Bartel. Their association would have to become publicly known sometime, he supposed, when she finally agreed to marry him. But maybe he'd manage to stop it; it wasn't really anybody else's business but theirs. He'd meant what he said, about preventing her any further hurt. It had been a mistake to cooperate on the damned book, irrespective of any association with Edward Hawkins.

At least the man hadn't pried about the corporation. Need he still tell Washington? wondered Patton. The rules were very explicit, because of the military and government contracts and particularly because of the front company proprietaries he had established for the CIA, for transportation to Latin America and the Gulf. But nothing had been said, to make him imagine Hawkins was even curious. Safe enough to leave it then. To tell them he'd need to refer to Chau Phu and he didn't want to talk about it any more. Today had been a reminder of something he'd long ago determined to forget. But failed. Because everything was so intertwined with it.

There'd been help – something else he'd tried to forget – but Chau Phu had made him known to the Army and Navy and Air Force Departments in those early years, after Vietnam, when he'd set up the business; made it easy, for them to put the contracts his way and for him to earn the trust which got the later approach from the Agency. Patton knew he had become rich – a millionaire several times over, in fact – because of that day at Chau Phu. And fallen in love with the wife of a victim. Patton didn't think he'd ever – in the secret, known-to-no-one-else part of his mind – be able to lose his feeling about that.

Chapter Six

Inside it was an uneven, surprising house. Hawkins followed Eleanor Peterson up a flight of stairs to a drawing room, from which French windows led out on to a patio upon which there were plants and trailing vines, with a canopied table and chairs as a centrepiece. It was floodlit, from some concealed lighting, and Hawkins thought, as he had before, that it looked like an illustration from a house beautiful magazine. So did the room into which she led him. Red, in varying shades, was the predominant colour, flocked wallpaper and heavy drapes and what he guessed was an Indian carpet, and which he hadn't seen on the previous visits. One wall was occupied entirely by a floor to ceiling bookcase in which the volumes were set haphazardly, some lying on top of others, as if it were frequently used. There were more books scattered around the room on low tables and even on the floor beside easy chairs and couches, which again took red as the colour. Near the library shelves was a grand piano, the top of which was clustered by photographs: there was no sharp overhead light, only sidelamps and the piano area was in the shadows but he knew the photographs: the majority were of Peterson with Party leaders. Two ex-presidents were in the picture gallery.

'I'm sorry about the confusion,' said the woman.

'It's my fault,' said Hawkins. 'I should have checked.'

She indicated a crowded drinks tray set near the verandah windows and said, 'There's most things.'

'Scotch,' chose Hawkins.

'Should I be doing this?'

'What?'

'Serving booze?'

'I'm sorry about last time.'

'You were a messy drunk,' she said and the open accusation surprised him: they were acquaintances, not sufficient friends for that. Was she going to mention the near pass?

'It's not a problem,' he said.

'If it weren't you wouldn't be calling it one.'

'You make it sound like therapy.'

'Maybe it is,' she said.

She poured, scotch for both of them, and as she handed him the glass she said, 'Cheers!'

'Cheers,' he said. Hawkins was surprised by her choice; he never thought of whisky as a woman's drink.

He sat in the chair she indicated and she sat opposite, on a couch which put her slightly lower than he was. Her blonde hair was loose about her shoulders, there was little make-up, and she wore just sweater and skirt and appeared far less formal than at the dedication. More relaxed, too, Hawkins thought; but then why shouldn't she? She was in her own home.

'Tonight was something John hadn't anticipated,' she said. 'A rally in Florida; a weekend opinion poll didn't show up very well there so he thought he'd better put in a personal appearance.'

'He's going to be very busy, now he's declared,' said Hawkins. 'Looking forward to it?'

'Is this an interview?' she asked cautiously.

'A conversation,' he said.

'Yes, I'm looking forward to it. I come from a political family, you know.'

'I do know,' said Hawkins. 'John told me.'

She smiled across at him. 'That interview caused a lot of flak. Political reporters here claimed the announcement should have been kept for American newspapers; some of his own people thought so too.'

Joe Rampallie would have been one of them, Hawkins decided. He said, 'I was very grateful.'

'John told me about your book.'

'I'm really only assembling material at the moment,' said Hawkins. 'I've a lot more things to sort out: more enquiries to make.' He tried to make it sound more difficult than it was going to be, wanting to impress her.

'Want to know something?' she asked suddenly.

'What?'

'When I left Vassar I wanted to be a correspondent,' she said. 'I majored in journalism. Then I did three years at Oxford, reading history. And in England I knew Edward Hawkins from what I read in the paper, long before I ever met him here, in Washington. Wasn't it the oddest coincidence that John should meet him in Vietnam, bringing everything together?'

'Very odd,' agreed Hawkins. 'Did you do any journalism?'

The woman shook her head. 'Which I regret, in a lot of ways. I came back here and got involved in politics: my father was up for re-election. I wrote a few freelance things, but nothing to set the world on fire. Then I met John and we got married . . .' she humped her shoulders '. . . and that was that.'

'I should think the White House is more attractive than the *Des Moines Courier*,' said Hawkins.

She smiled again and Hawkins thought she looked very attractive when she did so. 'There's a long way to go yet.'

'An incumbent has been upset before: and your husband's very popular.'

'Wasn't it one of your prime ministers who said something about a week being a long time in politics?'

Hawkins wondered who between Eleanor and John Peterson was the greater reader. 'Do you want it?' he said.

She looked at him mockingly. 'Can you imagine a woman who wouldn't!'

'I suppose not.'

There was a noise from the doorway and two children in dressing gowns giggled into the room, looking between their mother and the visitor. The boy was Asian. Eleanor hugged both of them and then said, 'Meet Mr Hawkins. He's a famous British journalist.'

Still giggling, the children came slightly forward and said 'Hello' almost at the same time: the girl was missing two front teeth and it came out as a lisp. A uniformed nanny followed the children into the room. Eleanor smiled up to her and said 'Thank you, Amy,' and the children were assembled and led, still sniggering, from the room.

'Your glass is empty,' said the woman, getting up from the couch and taking it from him.

'Sure I'm not keeping you from something?' He was beginning to feel warmed by the drink. Warmed but still in control. Which was how he had to stay.

'It was supposed to be a free evening, remember?' she said returning to him. 'I'm glad of the company,' she added.

'Why didn't you go to Florida?'

She sat down again, carefully arranging her skirt. 'It wasn't that sort of trip,' she said. 'This one's hotel meetings and friend-making, promises being made for favours to be returned. There's a lot of that to be done.'

Now it was Hawkins' turn to be mocking. 'Where did that cynicism come from!'

'I'm from a political family, don't forget,' she said. 'I know how things are arranged.'

'I'm not sure about the patronage system of American politics,' said Hawkins. 'It's always struck me as dangerous.'

'We copied it from the British,' she said. 'And what about your Honours List?'

'That's not the same,' disputed Hawkins. 'And we got rid of the worst sort of patronage at about the same time as you told George III he couldn't have America any more.' Hawkins realised he was enjoying himself.

'Deals go on everywhere,' she said.

'I meant to apologise,' he said. 'About last time.'

She shrugged. 'Those things happen: everyone's forgotten it now.'

'They were happening a bit too much,' he said. Why was he confessing to her? It seemed she'd only been conscious of the drunkenness.

'On top of it now?' she asked, in confirmation.

'I think so.'

'By yourself or with help?'

'By myself.'

'That's the tough way to do it.'

'Your husband didn't tell me that John was adopted,' said Hawkins, wanting to change the subject.

Eleanor held her glass in both hands, looking at him over the rim. 'Is this how it's done?' she said.

'I don't understand.'

'Interviewing . . . finding things out?'
'We decided this wasn't an interview; just conversation.'
'I'm not offended,' she said, hurriedly. 'I was just interested: wondering if I could have done it.'
'Could you?'
'I don't know,' she said. 'I suppose so, with practice.'
'Is this how it's done?' he said.
'I don't understand,' she said, entering into the game.
'Avoiding a question, with long political training.'
There was another smile. 'He's one of the babies John brought out of the orphanage at Chau Phu. He didn't have a name: few of them did because there weren't any records. John seemed the right name to call him. He's nine now. We keep April the twenty-fourth as his birthday: that's when he was born, to us.'
'I think that's nice,' said Hawkins.
'John junior is a worry in the campaign,' said Eleanor. 'I'm frightened newspapers will try to use him . . . use our adoption to create that sort of human interest crap they go for. Some people *in* the campaign are pressing for it.'
Rampallie again, thought Hawkins, remembering the staged meeting at the dedication. He hadn't expected Eleanor to use a word like crap: she didn't seem the sort of woman to swear.
'What's your husband say?'
'Absolutely no, of course!' she said, as if the question surprised her. 'The baby's out of it.'
'Does your husband talk about Vietnam?'
She shook her head. 'Why should he? It was just something that happened, something in which he became involved for a short time; a week, maybe a little more. It's been blown up out of all proportion, as if the only thing that John Peterson ever did was rescue fifteen kids from a Vietnamese orphanage. I know he saved their lives . . . no, I don't mean it like that. I know everyone involved saved their lives, but as far as John is concerned that's only one of the things he's done. He's been involved in a great deal more than that: done things as important if not more important.'
'I saw the film today in New York,' said Hawkins.
'What did you think of it?'
'It made me cry.' There was no embarrassment.

'A lot of men wouldn't have admitted that,' she said.

'I'm not sure why I did.' He'd relaxed completely with her.

'I'm glad you did,' she said. 'I think machismo is a pain in the ass.'

She'd relaxed with him, too, he realised. 'I met Eric Patton, as well.'

'You said you had a lot of material?'

'Trunkloads of it,' said Hawkins. 'Everything he ever did.'

'Here in Washington?'

Hawkins nodded.

'I'd like to look at it sometime,' said Eleanor.

'Whenever you like,' said Hawkins, surprised at the request. 'It's all there.'

'I'd really like that,' she repeated.

So, Hawkins decided, would he.

The main streets of Georgetown are garnished by restaurants and cafés, but Hawkins decided not to bother, although his earlier tiredness had gone. Instead he went straight back to Maryland Avenue. The only message being held by his answering service apart from Peterson's postponement was a dictated demand from Harry Jones for more chapters and queries on those already sent, and after checking and discarding his mail he went immediately to the basement. He'd followed his father's neatness in his cataloguing, creating his separate indexes but retaining the material in the individual trunks originally chosen by his father. Although, because of the man's meticulousness, it was unlikely that the CBS tape would have been misplaced in a wrongly-dated box, Hawkins started at the one that began in 1945 and worked carefully through, checking each one.

It was three hours before he was satisfied that it was nowhere among his father's possessions. He could have probably completed the search earlier but as he worked through he looked, too, for the notebook into which that morning's video had shown his father writing moments before the descent into Chau Phu. He couldn't find that, either.

Hawkins squatted back on his heels in the middle of the cases, an accustomed position now, staring from one to the other. He accepted that it was illogical, even impractical, for *everything* to have been retained. But Chau Phu had been important in his father's career: more awards and honours had come from it than from any other single incident. So he would have expected the man to have taken particular care of anything associated with it. And his father *had* been particular. So why wasn't it there?

Hawkins' legs began to cramp and he stood, climbing the stairs back up into the main house. Increasingly, he thought, he was coming up with more questions than he was answers.

The following day there were two communications.

The first was a letter from Colonel Elliott Blair. The man wrote that as required by the regulations of his service with the military he had submitted Hawkins' request for a meeting to his commanding officer. He very much regretted that permission had been refused, for unstated reasons. He wished Hawkins success with the project.

The second came within minutes of his entering his office at the press building. It was an invitation to interview the incumbent President of the United States of America, Nelson Harriman.

Chapter Seven

The last of the morning tourists were winding out on to Pennsylvania Avenue from the North door of the White House when Hawkins arrived. He went past and turned down in front of the Treasury Building, to enter through the East Gate. The authorisation was waiting and he was escorted by a Secret Service officer to the entrance to the East Wing Lobby where John Folger was already waiting. Hawkins was naturally accredited to the White House and attended the President's press conferences but had managed to establish nothing more than the strictest professional relationship with the Press Secretary. Now Folger strode out from beneath the portico, hand outstretched, and said enthusiastically, 'Glad you could make it, Ray,' as if the meeting depended upon the Englishman's convenience rather than the President's. Folger cupped his elbow and led him into the presidential residence along the East Wing lobby. As he moved along Hawkins saw the wood-panelled walls were lined with portraits of previous First Ladies and wondered if Eleanor Peterson's likeness would hang there one day.

'Familiar with your work,' said Folger. 'Gotta lot of respect for it.'

But particularly with the circulation of the Peterson interview, thought Hawkins. 'Thank you,' he said.

They went by what Hawkins presumed was the library and then a room in which varying designs of china were displayed in glass-fronted cases.

'How are you enjoying Washington?' asked Folger.

Hawkins accepted the man would know how long he'd been

in the capital from his accreditation application. 'Very much,' he said, small-talking. 'It's a fine city.'

Folger stood back to let him pass from the public to the working area of the White House. Almost at once Hawkins became aware of more people in the corridors. They turned left where a man was sitting at a table. He smiled up at Folger, then Hawkins, and led the way into what Hawkins recognised at once as the Oval Office. President Carter had installed the oak desk made from timbers of *HMS Resolution* and presented to America by Queen Victoria, and Nelson Harriman retained it. Positioned against the floor-to-ceiling bow windows overlooking the Rose Gardens Hawkins thought it looked very small: the one in Peterson's Senate office was nearly twice as big. Harriman rose from behind it, and came forward to meet him, gesturing to chairs set against a marble fireplace. There were vases with a Chinese design on the mantelpiece and above it hung a portrait of George Washington. The background was the American siege of Boston against the British.

'A pleasure to meet you, Mr Hawkins: a great pleasure,' said Harriman. He had a soft, almost melodic voice, blurred with a southern accent. He was physically a small man, white hair brushed back in wings from a clear, unlined forehead and with polished, shining cheeks. He wore a conservative, three-piece suit, blue with a slight grey overcheck and an unpatterned tie. The handshake was soft, a token gesture. Hawkins thought he looked like a senior bank clerk expecting one day to become chief teller but secretly dreaming of a managership.

'I'm very grateful to receive the invitation, Mr President,' repeated Hawkins.

There were chairs against the far wall and Folger discreetly settled himself. A witness, Hawkins recognised. He wondered if the tape system that had entrapped President Nixon during Watergate still operated.

'Didn't see the need for the formality of listed questions before we met,' said the President. 'There's nothing about which I'm unhappy to talk with you. On the record, of course.'

Harriman spoke looking towards Hawkins' tape machine.

Quite realistically Hawkins knew he had been granted the interview to provide a direct counter to that with Peterson, a response to what he had written. So the meeting was *his*, something that had happened without any association with his father. It gave him a good feeling: Harry Jones would have a hard time with any undermining campaign in London if he could carry on at this level.

'Ready when you are,' said Harriman.

Hawkins depressed the 'start' button and said: 'I recently had an extensive meeting with Senator Peterson, who has declared in the forthcoming election. Opinion polls assessing him specifically against you put him ahead, by a large margin. And there is a precedent in American history for incumbent Presidents failing for second terms. How worried are you by the Peterson challenge?'

From across the room there was a stir of discomfort from Folger at the directness of the question. Harriman gave his bank clerk's smile and said, 'I think it's always necessary to keep in mind that Senator Peterson has only yet declared, he hasn't secured his party's nomination. I am aware of Senator Peterson's popularity at the polls: he is a man of great charisma . . .' the man paused, to aim the shot '. . . but then there is always an appeal from a young man, irrespective of the proof of ability. And just as there is a precedent in American history of incumbent Presidents to lose a second term, there is an even greater precedent – common I would think to almost every President, whatever party – of latter-term unpopularity.'

'I would have thought Senator Peterson has proven ability,' pressed Hawkins. He realised the man had dodged the first question.

'Experience in the legislature is very different from experience in the Executive,' qualified Harriman. 'And making highly publicised visits to Europe and shaking hands on the doorsteps of official residences isn't exactly training for foreign affairs.'

Sensing that the other man had made a mistake, Hawkins said, 'It's in the realm of foreign affairs that Senator Peterson is most vocal. Particularly about your stance to the Soviet Union. How do you answer his accusations that you are over-conciliatory to Moscow?'

There was another smile from the President and Hawkins realised the man hadn't made a mistake at all: rather that he had guided the question, wanting to respond to it.

Harriman said, 'There has arisen a folklore about Senator Peterson that he is a man who has a deep experience of the horror of war: that he's been to hell and back. The fact is that Senator Peterson's combat experience is confined to a fortnight's visit to Saigon at the end of the Vietnamese war when he became involved in a particular incident, again highly publicised . . .'

He coughed and went on, 'I was at the tail-end of the Second World War; attached to Eisenhower's staff at Rheims. I saw, first hand, the devastation of Europe. I saw Dresden and I saw Berlin and I saw Cologne. I was on the staff of General Marshall when the plan was evolved to try to bring Europe from the wreckage of that war and I know the cost: not just the financial cost but the cost in human, physical suffering. I was in Korea, with MacArthur's headquarters staff, when America made its original mistake of imagining itself the world's policeman. And here, in Washington, during the sixties and the seventies, I saw the divisive effect upon this country of Vietnam . . .'

It was platform electioneering, Hawkins knew. But impressive platform electioneering.

'The definition of conciliatory isn't weakness,' continued Harriman. 'To conciliate is to gain and win over. So to the accusation that I am a conciliator I say yes, that I am. I want to gain the trust of the Soviet Union, not its distrust. To confront doesn't indicate strength. The reverse. To confront by building more missile systems and more nuclear submarines, *that's* weakness, the weakness of a bully threatening an opponent with a big brother unless he backs down . . .' Harriman hesitated again, for another point. 'You know what worries me about confrontation, Mr Hawkins? What worries me is the moment when, after you've confronted and confronted and produced even bigger and more horrifying weaponry than you had before, your opponent doesn't back down. That you have, to use the cliché, either got to put up or shut up. What happens then to your machismo? Do you imagine that the residents of New York or Washington or Buffalo or Los Angeles . . . or London or Paris or Bonn . . . are going to applaud belligerent confrontation then!'

'Senator Peterson's attitude is that confronted with determination, the Soviet Union will always back down.'

'Which is cras and naïve and not substantiated by one iota of evidence,' rejected Harriman, contemptuously. 'Krushchev was in an unwinnable situation when Kennedy confronted him over Cuba in 1962. And he was losing support within the Politburo, as well. To imagine Cuba as proof that Moscow will always pull back from the brink is to live in a dream-world.'

Hawkins was getting what would have made a powerful rebuttal on a debating platform, in immediate answer to what Peterson had said. But they weren't on a debating platform and so it wasn't immediate. The gap between the Peterson article and this would be at least two weeks, by which time the majority of his readers – certainly the English readers – would have forgotten what the Senator had said in the first place. Maybe he had the background for a feature but he didn't have a hard news line, apart from the fact that he had been invited by the American President to an on-the-record interview. He wanted more than that. He said, 'You withdrew your acceptance to an arms reduction summit after the improved SS-20 missiles were discovered?'

A satisfied expression settled on Harriman's face and he said, 'To achieve a point,' and stopped.

There was a moment of silence between the two men and a small surge of hope went through Hawkins. 'What point?' he said.

'Like so much else in recent months, identification of the new SS-20 system was projected as an embarrassment for me when, yet again, the reverse was true,' said Harriman. 'The embarrassment was Russia's. Which they knew and I knew and every sensible, experienced analyst knew . . .' He allowed a gap, for the repeated barb against Peterson to stab home. 'I could have reacted differently than I did, of course: gone beyond the withdrawal of my acceptance. But I wasn't interested in scoring cheap shots. I was more interested in achieving a summit than in wrecking it, by humiliating Moscow. I withdrew my acceptance but not the resident negotiating team in Geneva. From the chairman of that team I have got positive indication that the Soviet Union are prepared to accept reciprocal on-site inspection to ensure their guarantee and our guarantee of a halt to missile production.'

There was the familiar physical feeling of satisfaction as Hawkins recognised the importance of the disclosure. If Harriman could move aside the main barrier to every arms limitation debate, Peterson and any other presidential contender would be swept away like a leaf in a tidal wave. 'How strong is a positive indication?' he asked cautiously.

'The discussions have advanced sufficiently for a draft framework already to exist. It's not complete yet, but it's being formulated. When it's finalised, to the acceptance of myself and the Soviet Politburo, then the summit will take place and it will be signed.'

'Where?'

'That's yet to be decided,' said Harriman. 'Europe, I would expect: I'd be extremely happy to go there. Geneva has been a venue for such meetings in the past. So has Vienna. From either, of course, I would expect to visit London and fully acquaint your government of the details.'

He had it! thought Hawkins triumphantly. Harriman was using him, like Peterson – but while they gained their victories, he gained his. Cautious still he said, 'This is for attribution?'

'I've said nothing that I don't think can be quoted,' said Harriman. Cosmetically he looked across to Folger who made an uncertain gesture of agreement.

With no apparent limitation upon time Hawkins said, 'Can we turn now to domestic matters? Unemployment is extremely high and being blamed on the matchingly high interest rates imposed by your financial policies. Are there any immediate plans to reduce that unemployment?'

'The financial policies that I introduced upon gaining office have succeeded in reducing this country's inflation by three percent; the indicators as yet unpublished should show that in the immediate preceding month the drop is four percent. Inflation is a bugbear for a politician. People can see the price on a can of beans, or the length of a job line but they can't *see* inflation: its effect is on that can of beans, but that's not the way it registers. But it's something that has to be controlled . . .' The man gave another of his pauses and Hawkins waited expectantly, accustomed now to the man's manner. '. . . As I said a moment ago, the most recent indicators should show four percent. If that's the case then I don't see any reason why the Treasury Secretary shouldn't have discussions with the chairman of the Federal Reserve about some reduction in the discount rate.'

'By how much?' pressed Hawkins: he'd never expected a quarter of what he was getting from this interview.

'One percent certainly: if all other factors remain stable, it could be as high as two percent,' said the President. 'A reduction of that size would enable reinvestment and the pick-up of our industries. And if they pick up, so does the work force.'

Hawkins recognised the President had covered every point made by Peterson. And come out ahead every time. Folger stirred, from across the room, and Hawkins, satisfied now, took the hint. He said, 'I would like to thank you, Mr President, for the time you've allowed me.'

Harriman stood, terminating the meeting. 'Enjoyed it myself,' he said. 'Look forward to seeing you at the next open briefing.'

Hawkins filed behind Folger from the Oval Office but the man held back for them to walk level as they retraced their steps along the corridor, towards the East Wing. Folger was flushed and seemed more uncertain than he had earlier.

'There isn't going to be any briefing announcement, as far as I know,' disclosed the Press Secretary.

'I'd appreciate a call, if there is,' said Hawkins. 'I don't want to put this forward as something that it isn't.'

'Don't worry,' assured Folger, eagerly. 'I'll call straight away if there's any change in plan.'

'Thanks,' said Hawkins.

'And Ray?'

'What?'

'Anything you want now, you don't hesitate to call, OK?'

'OK,' said Hawkins, recognising an additional advantage from his meeting with Nelson Harriman.

Folger offered a card. 'That's my home number, as well as here.'

Harriman was anything but a bank-clerk President, decided Hawkins, revising his earlier opinion. The man was an adept, jugular-aware politician. Hawkins looked again at the paintings of the previous First Ladies as they approached the exit. Maybe Eleanor Peterson wouldn't feature here after all.

Eric Patton was a considerate lover, always concerned more with Sharon than with himself. He was tender too, the way he'd learned she liked love and she climaxed first and then again, when he did. His consideration stayed, even afterwards.

He didn't immediately withdraw and kept kissing her, bringing her gradually down.

'I love you,' she said.

'I wish I could believe that.'

'I've thought about it,' she said. 'I know you're right. I'm being stupid.'

He moved away from her in the darkness, trying to see her. 'About what?' he asked hopefully.

'I want to marry you.'

He snapped the light on and she flinched and turned away from the brightness.

'You really mean it?'

'Yes, I really mean it. But properly,' she said. She made a movement, encompassing the bed. 'I know it's stupid but I want to get married properly. Charles and I got married in a hurry, without any engagement and with only a week of his overseas embarkation leave as a honeymoon. This time I want an engagement. And time to plan the wedding.'

'You can have whatever you want,' said Patton. 'Anything.'

'I wish you wouldn't keep spoiling me.'

'I'll get a larger apartment in New York: up in the Sixties, where it's nice. Maybe Riverside,' he said, ignoring her remark.

She smiled gently up at him, happy at his boyish excitement. 'I don't want to move from here. I like it in this part of Connecticut.'

'Whatever you say: we'll have both because I want you with me all the time.'

Sharon turned to look directly at him, her face serious. 'That book you told me about: it couldn't embarrass us, could it?'

Patton leaned down, kissing her. 'Don't worry,' he said. 'He promised not to link us in it and if you want I can always ask to look at the manuscript: he did offer. Anyway, by the time it's printed, we'll be married, won't we?'

'I hope so,' said the woman.

Chapter Eight

Hawkins once again got the front page lead with Nelson Harriman's disclosure of the arms talks breakthrough and the accompanying feature, like that upon Peterson, ran across two inside pages. Both were copyrighted and because the White House only issued a brief, four-line confirmation, the syndication throughout America was greater than before. There was another flurry of congratulatory cables from London – even one from Harry Jones this time – and on the Monday Hawkins appeared on *Good Morning America.* There was the inevitable reference to his father but the thrust of the interview was his emergence as a leading political commentator in the American capital. Immediately afterwards he was approached by two different syndication agents offering to negotiate for him a weekly political column for US newspapers. He declined both. Apart from the possible conflict with his own newspaper he accepted realistically that what had happened with Harriman had occurred because of Peterson and that his could be a short-lived reputation.

But realistically again, he capitalised upon it in other ways.

Until now his relationship with Jeremy Abbott, the press counsellor at the British embassy, had been at about the same level as that with Folger. The day after the appearance of his exclusive meeting with the President Abbott extended the invitation for a meeting with the ambassador. Sir Neville Wilkinson was a tall, powerfully-built man with a port-mottled face and a monocle in his left eye, which Hawkins couldn't determine as an affectation or an indication of a flamboyant character. The ambassador used clichés like a useful exchange of information and said he frequently gave off-the-record briefings to selected British correspondents and that he would be glad if Hawkins would consider himself one of those special few. Hawkins recognised that Wilkinson wanted to receive more than he gave but he welcomed the level of contact and said he appreciated the confidence the ambassador was extending to him.

He accepted, too, a luncheon invitation from the Speaker of the House, Paul Brunton, and had meetings, at their suggestion, with the majority and minority leaders of both the Senate and the House of Representatives. They gave him a consensus of political reaction to what Harriman had said and provided another major feature for the following Sunday.

It was a busy period – busier than he had ever known – and behind the professional satisfaction there was a personal one. He only got drunk once. He did it quietly, at Maryland Avenue, a solitary celebration that no one else knew about and the following day, without any difficulty, he didn't have a drink at all, which increased the personal satisfaction even more.

He was too occupied to contact Peterson about the postponed meeting and before he was able to, the approach came from the senator, through Rampallie. This time he caught a cab to the Senate building and discerned at once a change in the attitude of Rampallie towards him. Before, Hawkins had been aware of a vague condescension but now Rampallie behaved more like Folger had done after his favoured treatment by the President. As they approached Peterson's office Rampallie confided, 'More time today: the senator's schedule is looser than before.'

On political balance Harriman had won overwhelmingly in the opening exchanges and Hawkins entered Peterson's office curious of what the senator's attitude would be. Peterson strode smiling across to meet him, as he had on the earlier occasion and as before, gestured him to the worn leather furniture along the wall.

'Been hearing and seeing a lot about you lately, Ray,' he said. There was no rancour in his voice.

'It's been a good time, professionally,' said Hawkins.

'Harriman creamed me,' admitted the politician at once.

Hawkins hadn't quite expected such openness. He said, 'Yes, I guess he did.'

'That's the advantage an incumbent always has,' said the senator. 'We can say what we will do, but the President can actually make things happen. It's a sonofabitch but we've got to be philosophical about it.'

'How are the polls showing, nationwide?'

'For Harriman, every one. Couldn't expect anything otherwise. We've got inside track of his platform – "Peace and Prosperity" . . .' Peterson shrugged. 'And that's a pretty damned good platform.'

'You don't seem depressed about it.'

'If I suffered from depression I wouldn't be in politics,' said Peterson. 'There's a long way to go, before the election. Employment's got to be seen to go down, because of the interest rate cut. And the Geneva talks haven't actually produced the framework yet.'

'That's being objective as well as philosophical,' said Hawkins.

'You coming to the convention?'

Until now Hawkins hadn't decided but he said, 'Yes, of course.'

'Ever been to Miami Beach?'

'No,' said Hawkins.

'I'm staying at the Fontainebleau. I'll see that Joe arranges all your accreditation.'

'Maybe I could travel with you too?' He'd have to make similar arrangements with Harriman. Traditionally Presidents didn't stump the country as much for their second term so Peterson was the major priority.

'Whenever you want,' agreed Peterson. 'I'll see Joe fixes that, too.'

Coming to the purpose of the meeting Hawkins said, 'I went up to New York. Saw Harvey Lind's film and met Eric Patton again.'

'How can I help you further?'

'By telling me what it was like, as you saw it.'

Peterson had been looking directly at the other man. Now he stared down reflectively and said, 'You know, my feelings about that are more mixed than about almost anything else. Because of what we did that day fifteen kids are now alive and loved. One is my own and I love him so much it hurts. But seven Americans are dead: Americans who would be alive and with their families today if I hadn't insisted on going to Chau Phu.'

'I don't know if there is an equation,' said Hawkins. He was constantly surprised by Peterson's honesty.

'No,' said Peterson. 'I don't suppose there is.'

'Patton said he thought they were careless, in not posting pickets.'

'Enquiries have to find guilt; that's their function. You've seen the film. Those kids were dying. Forest had done a recon, satisfied himself. OK, upon hindsight, there *should* have been pickets. But at the time it seemed more sensible to get the kids into the helicopter and get to hell out of it . . .' He stopped and then said, 'If blame's important, then I'm the one it should be apportioned against. I'm the one who yelled at everyone to come and help.'

'How did you feel, at the time?'

'Frightened as hell. Not at once. There wasn't time, at first. Initially I just reacted, did what Blair said and ran after him. The fear came when I was lying over the kids. It's funny, how your mind works at times like that. There I was, thinking I was going to die and all the time I was trying to raise myself just slightly, so that I didn't crush the children. That was the moment I decided that if I got out I wanted John. He was up against my right shoulder. He had lice and he had scabies and an open wound on his right arm: don't know how he got it although the surgeons back here said it looked like rat bites, as if they were gnawing at him. Some of the kids were so ill they didn't react to the firing, but John did. He kept jumping but then there was so much incoming it developed into a perpetual shake. I said to myself that if I got out I would adopt that child and make it so he never had to shake again in his life . . .' Still looking down in reflection Peterson shook his head, at internal sadness. 'It hasn't really worked. That kid's surrounded by so much protection he's safer than Fort Knox and last weekend I took him to the kite shop in Georgetown and a car backfired: he went down into a crouch, whimpering, faster than a cat off a roof.'

'What about my father?'

Peterson looked back and said, 'Want to know something else? I was so scared in those first few moments that I didn't even think about anyone other than myself. It was only in the helicopter that I thought of other people. Blair was on the gun in front of me and I looked around, expecting to see the others. But there was only your father. I couldn't see where they were, because of how I was lying, but I realised they were still out there, trapped. I don't think your father was as frightened as I was. Guess he'd seen too many wars: was too used to it. Shells were actually ricochetting off the metal-work of the helicopter and he was trying to pull kids underneath him, like he was making them tidy. One kid's leg was hanging out of the machine. Your father pulled it inside and within minutes something hit the very spot where it had been. God knows what it was, a shell of some kind obviously. It would have blown that kid's leg away. We didn't have any medics aboard, so he'd have died, if it hadn't been for your father.'

'You make him sound very brave.'

'I suppose I do,' agreed Peterson. 'But that isn't the impression I retain of him. It was . . .' There was another hesitation, while the man thought for the right expression. He began again '. . . like he was used to it. There we were, caught in an ambush that I never thought we'd get out of and your father was calmly collecting kids together, getting them out of danger, shielding them. I felt a fraud, lying there. I wasn't doing anything. I wasn't working the gun, like Blair. Or the rifle, like Patton. I was as close to the floor as I could get, without suffocating the kids, but your father was propped up on his elbow, collecting them altogether.'

There was none of this from the film or the talk with Patton or from his father's restrained, unemotional account, remembered Hawkins. He thought again of how he intended to begin the book – My father was a hero, a man of whom I'm proud. It was very fitting, he decided. 'How long did it last?' he said.

'Like forever,' replied Peterson. 'Then, suddenly, Patton took us up. I thought we were all going to fall out. The machine was open-sided, remember, and none of us were strapped in. But it didn't happen; something to do with the centrifugal force, I suppose. We tipped over: I was looking through the doorway through which Blair was firing and I actually saw them. I actually saw the guys trying to shoot us: up close, not as close as we are now but close enough to see what they were wearing and how their faces were twisted and how they were firing. It was like a see-saw: made me feel sick, although I don't know if that was fear or because of the way the helicopter was moving. We went one way and then we came back, to go over where the others were.'

'Did you see them?'

Peterson nodded, not immediately replying. 'Patton banked,' he said at last, 'a lot of times. Six, seven passes at least. I was looking down at them like I'd been looking down at the Vietcong. They were all gone. Dead.'

'What was said?'

Peterson shook his head. 'Nothing: nothing that I can remember anyway. I guess there must have been something but it wasn't much. Everybody was inside himself, thinking about what had happened.'

'What about Lind?'

'Blair tried to treat him: see what the injuries were. It was hopeless. He'd taken several in the chest: God knows how many. Half his head was gone, as well. He must have died before we even left the ground.'

'You've said you were scared?'

'I was,' said Peterson.

'Do you mind my saying so?'

'Why should I? It's the truth.'

'I just wanted to get it straight.'

'That's how it happened; that's how I felt.'

Hawkins decided he had all he wanted for the biography. He said, 'Maybe it's too early at the moment but perhaps I could come along during an early stage of the campaign, after the official endorsement?' The direct comparison between Peterson and Harriman had been thrust upon him but Hawkins thought suddenly that it would be an interesting theme to continue, throughout his coverage of the election build-up. It would be a comparatively easy concept to carry through if Harriman allowed the cooperation that Peterson was offering: and the President had indicated through Folger that he was prepared to do just that.

Peterson appeared relieved that Hawkins had moved the conversation on from the Vietnam reminiscence. He said, 'Like I told you, come aboard whenever you feel like it.'

Hawkins stood and the politician said, 'By the way, Eleanor has told me about your father's records: said she would like to look at the cuttings.'

'Any time,' said Hawkins.
Eleanor Peterson telephoned that night.

She wore jeans and a sweatshirt and no make-up and Hawkins thought she looked like a college girl when she arrived at Maryland Avenue. A beautiful college girl.

'I had to have John ask you,' she said, immediately she entered the house. 'After suggesting it that night I got embarrassed: thought you might think I was silly or something.'

'Why should I think that?'

The uncertain shrug was schoolgirlish, too. 'I don't know. Sure I'm not intruding?'

'Quite sure.'

He led her over the echoing hallway into the disordered study, where he'd built the fire up: that morning there had been the first snow, sudden, gale-driven flurries. She crossed to it, shivering, and stood with her back to the flames. Nodding to the desk and the open typewriter she said, 'How's it going?'

'Chapter seven,' he said.

'Happy with it?'

'I think so.'

'When do you think you'll be through?'

He made an uncertain gesture. 'Few months yet.' He looked towards the drinks tray and said, 'Would you like a drink?'

'Would you?' she said pointedly.

'No.'

'Later then,' she said. 'Not now.'

'OK.'

'Saw you on television,' she said.

'What did you think?'

'I didn't think you made the comparison between the monetary policies of Great Britain and America convincingly enough,' she said. 'And your shirt collar stuck up over the edge of your jacket.'

Honesty appeared to be a trait of the family, thought Hawkins. She was right about the economic question: it was the one with which he'd had most difficulty. He hadn't known about the shirt collar. 'Thanks!' he said.

She looked at him, surprised. 'You asked me!' she said.

'You could have lied.'

'You were terrific.'

'That's better.' Hawkins was uncomfortably aware of the flirtation. He said, 'All the cuttings books are in the basement.'

She seemed aware of it too. 'I'd better go down,' she said. 'I want to get home in time to see the children before they go to bed.'

He led her down the stairs, conscious of the untidiness when he got into the basement. 'I've been working on them,' he apologised.

'I'll try not to disturb anything,' she said.

Hawkins explained the progression of the trunks, from the beginning until the end of his father's career, pulled the chair closer for her and arranged the anglepoise light. 'OK?' he said.

'Would you be working down here now if I wasn't here?' she said.

'Probably.'

'So I'm in the way.'

'Don't be silly.'

'Why don't you work here anyway? Would I disturb you?'

You'd disturb me like hell, as you're doing now, he thought. Aloud he said, 'I've got things to do upstairs. I'll come down later.'

Back in the study he read the *Washington Post* and *Newsday* more thoroughly than he had earlier, clipping items he thought might be expanded into more comprehensive stories for England, conscious all the while of the woman below in the basement. *Had* she realised the flirtation? And been unoffended by it? Of course not. She was just a friendly, open girl. It was madness to imagine anything more: fantasising madness. Eleanor Peterson was the wife of a man running for the presidency of the United States of America, happily married and the contented mother of two kids. So what the hell was he thinking about? He wasn't even drunk!

Remembering her wish to get home before the children were put to bed, Hawkins moved towards the basement after two hours and encountered her halfway up the stairs.

'I thought you might have forgotten the time,' he said.

'I almost did,' she said. 'I know what you mean about there being so much material. Isn't it a shame you won't be able to talk to Ninh?'

'My father's interpreter?' frowned Hawkins. His father had

often talked about Nguyen van Ninh during their conversations.

Eleanor finished climbing the stairs and he stood back for her to get into the hallway. 'I guessed that was what he was, from the references in the notebooks. I started like I usually read books, at the end first. The last trunk, with all the clips about Vietnam.'

Hawkins had invited her only to read the newspaper cuttings; he felt a stir of irritation at the intrusion. 'Yes,' he said. 'I suppose it is.'

'Do you know what became of him?'

'No,' said Hawkins.

'I wonder if he got out? Or if he's still there?'

'I would have thought he would have attempted to make some contact with my father, if he got out.'

'Do you know what struck me, about all the stories?'

'What?'

'He was *there*. He was always describing the things he'd seen, that he'd witnessed. Or quoting directly people who'd been involved. I think it's wonderful that a man as renowned and famous as he was still took notes in a book held together with a metal spiral at the top, just like in the movies.'

'It was a principle of his,' said Hawkins. 'He didn't like getting things secondhand.'

'I didn't get through one trunk,' she said.

'Come back any time,' offered Hawkins. He wondered how to prevent her looking through the notebooks and said, 'But I've indexed a lot of stuff: I'd appreciate it staying as it is. The cuttings books are fine.'

'I'll be very careful,' she promised, and he knew she'd missed the objection.

He went with her to the door and she continued on when he expected her to stop, so that when he reached across to open it for her they came close together, their faces only inches apart. Momentarily they remained that way, serious faced, one looking directly at the other.

'We didn't have that drink,' he said.

'Next time,' she said.

Chapter Nine

Hawkins was confused by Eleanor Peterson. Had it been a challenge during those brief seconds in the doorway? He couldn't decide. Nor did he want to. Whether or not John Peterson made President was immaterial; she was still the wife of a very important politician in a city where politicians were everything. And powerful. Attractive though he found her – and he found her worryingly attractive – it would be utter insanity to let dividing lines become blurred. He knew he was too sensible to allow that to happen. And that she was, too. So it was stupid to build a few seconds of uncertainty into a problem that didn't exist. Still maybe wise during her next visit consciously to keep a distance between them.

Hawkins went down into the basement and sat in the chair in which she had sat and was aware of the lingering traces of her perfume. She'd pulled the last trunk nearer, for easier access, and although it meant jumping from his methodical indexing he reached in, not for the cuttings books but for the notebooks. He read steadily, browsing, caught as she had been by the references to Nguyen Ninh.

He'd described him to Eleanor as his father's interpreter but Hawkins knew the man to have been more than that. Officially Ninh had been manager of the office his father had described to him, near the cathedral at the top of Tu Do Street. He had been the man who arranged and then kept up to date his father's accreditation at the time-consuming Military Assistance Command headquarters. He had been the man who arranged and then kept up to date his father's accreditation at the time-consuming Military Assistance Command headquarters. He had been the man who knew how much and who to bribe at Tan Son Knut to get airfreight packages en route to London. He knew the right people in the black market that flourished in the road parallel to Tu Do and the name of which Hawkins could not remember, so that for parties he could buy bottles of gin that had gin in them and not kerosine pumped in through the drainage hole in the bottom made with a hot wire. And he knew which of the traders were Vietcong and which were not, so that when he bought the candles for the frequent power failure black-outs he got those made throughout with wax and not just coated around the explosive for which the wick acted as a fuse.

Eleanor had been right; it was unfortunate that he wouldn't be able to talk to Ninh. Something else she'd said about his father came into his mind. '*. . . he was there . . . quoting directly people who'd been involved . . .*'

Ninh *had* been involved, in the last active years of his father's career. Probably more than anyone else; the person who saw him daily, spoke to him daily.

Hawkins sat back in his chair, staring unfocussed at the Vietnam trunk. Why not? London would agree: probably even encourage it. People had got back, since the end of the war, on tourist visas through Hanoi. Objectively Hawkins accepted that the possibility of finding Nguyen van Ninh was slight. The man could have died or fled or disappeared in a hundred different ways. But there were reasons for going, beyond trying to trace Ninh. Hawkins had seen the films and the photographs and read the stories and heard the anecdotes but he still didn't *know* what it was like. He hadn't experienced the heat and smelt the smells: hadn't *felt* what it was like.

London were as enthusiastic as he anticipated, and Hawkins filed the application through a travel agency on Connecticut Avenue that took more than an hour to complete. Joe Rampallie contacted him the same day, offering full facilities whenever he wanted to accompany John Peterson on his campaign. They had taken a block booking of accommodation at Miami Beach and there was a room available for Hawkins if he wanted it during the convention. Hawkins thanked him and accepted it and the following day received a package that must have weighed at least two pounds, containing all the documentation he would need throughout the party conference and confirmation of all the arrangements they had made. In addition he saw that he had been allocated a booth and telex facilities in the press room.

Hawkins repeated the process with the White House and found John Folger as helpful as the press secretary had promised to be after the meeting with Harriman. As Hawkins anticipated, the President did not intend as much campaigning as his chief rival but Hawkins was accredited to the trips already arranged and Folger undertook automatically to add his name to those which were decided upon later. Their convention was in New York and the President intended staying at the Waldorf Astoria. Folger told Hawkins to wait an hour from the end of their call before making his request for accommodation and when he did received return confirmation of his reservation. The press kit he received from Folger was lighter than that of Peterson but just as extensive. Again he had been allocated an unshared booth and telex facilities. The day after their initial conversation Folger called back to say if Hawkins advised a week or two in advance which one he preferred he would arrange for the Englishman actually to accompany the President aboard Air Force One on one of the trips he was undertaking.

It was a week before there was any contact from Eleanor Peterson. This time she arrived at Maryland Avenue earlier than on the first occasion, only minutes after Hawkins returned there himself from the office.

'There's so much to read,' she said, as if offering an explanation. She again wore jeans, with a fur jacket over a wool shirt and a woollen hat pulled low to cover her ears. Her nose was red from the cold.

'My father would be flattered by the interest,' said Hawkins. He decided upon a positive effort to remain aloof.

In the study she stood close to the fire again. 'I'm taking the children to the Caribbean next week,' she said. 'St Lucia. Have you been there?'

Hawkins shook his head. 'The sun will make a change,' he said.

'Little John needs it: he feels the cold dreadfully.'

'I fixed up to accompany your husband during some of the campaigning. And I'm coming to Miami Beach,' Hawkins said. He indicated the drinks and this time she nodded agreement.

'That's the part of American politics I don't like,' she said. 'All the glad-handing and theatricals. I think it's demeaning: it embarrasses me.'

'I would have thought you would have been used to it.'

'Nope,' she said. 'Never been able to adjust.' She accepted her glass and held it before her, looking at it. 'I know what it's like. This I mean.'

'What?' frowned Hawkins.

'John thinks I drink too much.'

'What do you think?' said Hawkins, embarrassed by the confession.

'That he shouldn't worry until I let him down. Which I won't. You know what it's like – it helps, that's all.'

'Helps what?'

She didn't reply at once, turning down the corners of her mouth. 'All sorts of things,' she said. 'Boredom. Meeting people. Something to do.'

'That sounds familiar,' said Hawkins.

'I told you a lie,' she said. 'Well, not a lie exactly; just not the whole truth.'

'About what?'

'Going into journalism, when I left university. I didn't try to get a job. I was frightened.'

'Of what?'

'People. Having to make approaches all the time to complete strangers: having to get to know them and get information. I've always recognised my shyness. That's why it was better working with my father. I knew the people and I knew the Washington routine: nothing was new or unsettling. It was comfortable: safe.'

'How does that reconcile with being the wife of a man running for the presidency!' said Hawkins. He felt sorry for her.

She stared down into her empty glass. 'It doesn't,' she said. 'Not very well anyway. That's why I don't go around with him much: just the special occasions, when it's absolutely necessary.'

'You shouldn't be telling me this,' said Hawkins.

She offered her glass and as he refilled it she said, 'You wouldn't write about it, would you?'

He came back, looking directly at her. 'No,' he said. 'I won't write about it. But I don't think you should have put me into the position of having to make a promise.'

'Sorry,' she said. 'I always seem to be apologising, don't I.'

'Stop it!' he said.

'What?'

'Parading your inferiority complex.'

'You wouldn't have known what it was, if I hadn't told you.'

'But you did tell me,' he said. 'You've no need to keep apologising here.'

She answered his direct look and after several moments said, 'Don't I?'

'You came to look at the files,' he said, moving back.

'Yes,' she agreed. 'I did. This time I'm going to do it properly; from the beginning.'

'What time do you have to be back?'

'John's away: Philadelphia,' she said. 'And I saw the children before I left.'

'I'm grilling steaks,' he said. 'There's clam chowder but it'll be out of a can.'

'You inviting me for dinner?'

'Yes,' said Hawkins. Stop being a bloody fool, he thought.

She held his eyes again and said, 'I'd like that very much.'

Hawkins laid a table in the kitchen. The steaks were fillets, so there was no need for them to be tenderised. He made the salad and then his own dressing and pulled the Brouilly to let the wine breathe, trying to occupy himself with activity to avoid answering his own doubts. If he stretched much further towards the fire one of them was going to get burned, he thought. What had happened to the determination to remain aloof? She emerged after two hours, still wearing reading glasses: they made her look serious. She had a smudge of dust on her cheek.

'I got through the first trunk,' she announced. She stretched back to ease the cramp from her shoulders and Hawkins wished she hadn't because it showed how heavy breasted she was.

'You've got dirt on your face.'

'I've got dirt all over me! Where's the bathroom?'

'To the left, in the hall.'

He'd warmed the soup by the time she returned and put the

steaks beneath the grill. He touched his wineglass to hers and said 'cheers' and she said 'lucky life' in return and after tasting the soup said, 'I couldn't tell the difference.'

'No,' agreed Hawkins. 'I'm never able to, either.'

'Why haven't you got married?' she said, in another of her abrupt questions.

'How do you know I'm not?'

She stopped drinking the soup. 'Are you?'

'Was,' he said. 'Her name was Jane and she had blonde hair, not unlike yours.' Was that what caused the initial attraction, the similarity? It hadn't occurred to him before.

'What happened?'

'It was a university romance,' said Hawkins. 'Cheap wine out of plastic cups, everything mattering very much and seeming more important than it was.' Had he been looking for a mother rather than a wife, like Jane had said over and over at the end, when she'd started to hate?

'What about kids?'

Only the one she was expecting and lost after seven months, the reason for their marrying in the first place, Hawkins remembered. 'Fortunately not,' he said.

'Still keep in touch?'

'There didn't seem any point,' he said. Which wasn't the truth. She'd despised him in the end, calling him ineffectual and weak and even impotent, which he'd become because of the rows. He'd tried, with letters and cards but she'd never responded, not once. He stood up and went to the grill and said, 'How do you like your steak?'

'Medium,' she said. 'Think you'll try it again?'

'I haven't really thought about it,' he said. Which was another lie. Hawkins knew he'd be frightened of trying again, of making another mistake. He'd hated the rows and the accusations.

He served the meat and she said, 'Shame. You'd make some wife a great cook.'

'There's not much you can do wrong with steak.'

'You should see the way I cook them! If we didn't have staff we'd starve.'

Afterwards he offered fruit and cheese but she didn't want either so he suggested they take the remainder of the wine back into the study, where the open fire was. He waited until she chose the settee and then carefully sat in an adjoining armchair.

'I'm trying to get to Vietnam,' he announced. 'It was the remark you made last time about Ninh. Don't know if I'll be lucky or not.'

'He might not be there.'

'It'll still be worth the trip.'

'How long will you be away?'

Hawkins shrugged. 'As long as they'll give me a visa for, I suppose.'

'Are you going to kiss me?'

To cover the instinctive swallow of nervousness Hawkins drank some wine. He said, 'For someone who's shy, you've got a hell of a way of showing it.'

'It's a trick,' she said. 'That's how I try to cover it, by appearing to be the other way.'

'Do you want me to?'

'Yes.'

'Do you think it's a good idea?'

'No.'

He moved over to the settee and took her glass from her and kissed her, awkwardly at first because they were both nervous and then more comfortably, as they settled against each other.

'I'm not going to sleep with you,' she said.

Unable to stop himself, even though he was so close to her, he laughed. 'Christ, you're the strangest woman!' he said.

'Do you want to sleep with me?'

'I don't know,' he said. 'I think so, yes. Yes, of course I do.'

'You don't have to pretend,' she said. 'That's how I feel, I think so but I don't know. I won't sleep with you until I'm sure.'

'What happens if you can't make up your mind?'

'Then we won't.'

The following day Hawkins' visa application to visit Vietnam was authorised from Hanoi. Apart from the professional advantages he decided it was probably a good idea he was getting out of Washington for some time.

Chapter Ten

Impressions crowded in upon Hawkins and later, in Washington, he was to be glad he kept a detailed diary. The first entry was of the heat, an enveloping, breath-sucking heat. He numbered the agricultural developments they toured – 'six, all the same' – and recorded that the land was extremely flat, and that the peasants did wear hats like lampshades. There were a lot of irrigation dykes and he wondered whether they were the ones the Americans held back from bombing or whether they were nearer the delta mouth, at Haiphong. There was an entry which said 'everywhere bicycles' and that the war museums were predictable. The buildings in Hanoi remained predominantly French colonial but 'slowly crumbling'.

There were a lot of Indians and Asians on the tour and eight Europeans. Three were Swedes from a Socialist college near Stockholm, another an East German, a Frenchman, two affectionate middle-aged Englishwomen who constantly held hands and whom it was discovered halfway through were sisters and an angular, bespectacled girl from Düsseldorf named Heidi Becker to whom Hawkins loaned his personal bath plug – a remembered tip from his father – and who made it clear she'd be fulsome with her thanks if he came to her room to collect it. She accepted his refusal with sad but accustomed resignation.

Their guide was Mr Tho who saw his function as filling every waking moment of their time with activity and who rebuked Hawkins for wandering away from official hotels without informing him. The first time, in Hanoi, Hawkins encountered a group of Russians in a bar. He watched for an hour while they got miserably drunk and recorded it in the diary alongside the number of Russian and East European ships he counted in Haiphong harbour – eight – and the preponderance of East European goods in the shop windows.

On the fifth day they went south.

Hawkins let the rest of the group precede him at Tan Son Knut, lingering between the aircraft and the surprisingly small terminal building, trying to create an atmosphere. Had it been from this spot that his father had taken off and landed, that last time? And so many other times? Or somewhere else, somewhere he couldn't even see in the huge vastness of the complex. There were quite a lot of aircraft at the airport; many were unmarked but he identified some with Pakistani and North Korean designations; to his left the regimented helicopter pounds, line upon line of box-like enclosures, were still in place but there were few helicopters now. Mostly the spaces were occupied by airport vehicles.

As soon as they left the immediate airport road, the suburbs came out to meet them: tall, storeyed buildings, some several floors high, interspersed with low ramshackle constructions, with sagging corrugated roofs and haphazardly tacked wooden fronts. There seemed more people than there had been in Hanoi. The roads and the sidewalks were crowded and the highways were thronged, motor-scooters upon which pillion passengers perched unsafely side-saddle weaving and darting between the larger traffic, like worrying wasps. Despite fading and the attempts to obliterate the words, it was still possible to see written traces of the American presence. Hawkins read 'Florida' across the front of a darkened, shuttered building he assumed to be a bar and further on, across the top of a café that was open, the word 'Flamingo' and then 'Dancing'. There had been a word before but that had been heavily painted through. They crossed a river along the banks of which Hawkins got a fleeting impression of a wooden shanty town, board hovels appearing to stick one atop the other and against the very slope of the banks like birds' nests, mocking gravity.

And then the highway broadened out, along the routes planned and created between shading plane trees and spacious, set-back villas not by the Americans but by the French who had preceded them. In the bleaching sunlight of the day Hawkins guessed the impression would be difficult but now it was easy to imagine this as a suburb of any small French town.

Abruptly, to his right, a large parkland formed into a square and then Hawkins saw the huge red cathedral and knew where he was. He strained, trying to locate the presidential palace on the far side of the park, through the gates of which the North Vietnamese had driven their tanks with flowers protruding from the muzzles, but it was too dark now.

The buses went past the cathedral to the left and Hawkins knew he was in Tu Do, the main street of what he still thought of as Saigon, despite Tho's constant repetition of Ho Chi Minh city. Images kaleidoscoped in upon him. Walled and balustraded barracks, another small park fronting a building which escalated upwards in serried ridges until it became at least fifteen floors high, bars and shops to his left and then the place he'd had described to him so often and which he'd tried to reconstruct in his mind, just as frequently.

The buses stopped there and Hawkins disembarked with the same reflective slowness with which he'd left the aircraft.

The coaches had pulled off the main road. To his right the open-sided verandah room of the Continental Hotel was exactly as he'd thought it would be, with the diminutive bar against the wall at the rear and arranged before it, as if for inspection, the small circular tables attended by sagging rattan chairs. On the other side of the road in which he stood was the red-roofed Assembly building, disappointingly small for what had once purported to be a parliament and in front of which saffron-garbed monks had immolated themselves in religious protest.

And on the far side of the Assembly was the other main hotel of the city.

His father had made his permanent home at the Caravelle. The man had always had the same room, at the side overlooking the Assembly building, away from the blasts of the rockets which the Vietcong lobbed from the other side of the river with devastating accuracy, like giant footsteps, up the main street. When his father went on his rare vacation trips to Singapore or Hong Kong the management simply locked up his room and held it for his return, maintaining the sticking plaster against the windows to prevent splintering from a stray explosion, storing his possessions in his reserved and secure locker in the basement. From its flat roof, after curfew, his father had stood with other correspondents and watched the red-streaked firefights: it had actually been possible, with the properly-tuned radio, to eavesdrop on the conversations between Ton Son Knut control tower and the helicopter gunships and the low flying Phantoms. Everyone had had properly-tuned radios.

Directly in front of the Assembly was an open space not big enough to be called a park. Here, Hawkins knew, there had once been the statue intended to commemorate the comradeship and cooperation between the American and South Vietnamese forces, two huge carved soldiers, one behind the other. One definition had it that the rear figure was the South Vietnamese soldier, hiding behind the American; the alternative was that the rear soldier was the American, urging the South Vietnamese into battle at bayonet point. The café his father had described still existed, on the opposite corner, and Hawkins wondered if the French-style delicatessen remained behind it. It was there the correspondents bought their processed meats and imported cheese and stick loaves and Chablis and Beaujolais in ring-topped cans, before catching a cab to watch the war. Hawkins thought a photograph of a front-line picnic, with tins of wine in cooler packs, would have made a good exhibit in the war museums he'd obediently toured in the north. Another snide reflection: he didn't know so how could he sit in judgment?

Hawkins feigned illness, and the need to remain constantly near a bathroom, to avoid the perpetual activity of Mr Tho. And tried to discover the Saigon his father would have known. He breakfasted in the open garden of the Continental where his father began every day, under dull-leafed trees, with scavenging cats in their roots and unafraid birds in the branches, contestants for the crumbs. He drank beer – but not much because it tasted of onions and he didn't have the need for booze anyway – on the open terrace of the hotel, served by an ancient man with a carved, unmoving face, who had probably waited on French, then Japanese, then French again, then Americans and now conquerors from the north. He looked for but could not find the bar his father had told him Graham Greene used when he came to the city to write *The Quiet American* and where the writer met the girl upon whom Phuong was based. And he walked up Tu Do and sat in the massive, red-brick cathedral, oddly out of place, as if God had put it down for a moment and forgotten about it and which was hot inside, not cold as churches were supposed to be.

And he tried to find Nguyen van Ninh. It took a long time. He had an address from a fading American military form accrediting Ninh as his father's interpreter, with the right to attend South Vietnamese army briefings, but without a street map it was meaningless. From the graven-faced waiter at the Continental he was gestured vaguely towards the docks and waited until the heat started to go out of the day before setting off: even so he felt the wetness gather almost at once, glueing his shirt to him.

The waterfront stretched along the broad sweep that later became the Mekong, with a straggle of ocean boats and river craft against the bank and at anchorage further out, all watched over by droop-headed cranes. Four of the bigger ships were Russians and two were East German. Hawkins wandered parallel to the river, gradually aware of the shacks and lean-to huts of a shanty ahead of him. It was at the join of a small tributary into the main river, a thread of water so narrow that Hawkins couldn't decide if it were a proper, minor river or an expanded drainage ditch.

A lot of the patchboard buildings were raised on uncertain stilts or odd pieces of stone – against flooding, Hawkins presumed – and children skittered and played in the gaps. There was a stink of sewage that caught in his throat, making him swallow. Closer, Hawkins was able to see that the slum cluster wasn't as haphazard as he had first imagined it. Overwhelmed by debris and sag-sided buildings was what had once been one of the straight, well-built French thoroughfares. There were even the shells of colonial villas, three window-gaped houses forlornly lost amidst a tangle of neglected garden undergrowth. Children started emerging from beneath the houses, to stare and giggle at him curiously: three of the braver ones formed a group and began to follow him. He turned and said 'Hello' and they scattered back to safety, giggling even more.

There was no gate to the first villa but its long-ago supports remained, gradually collapsing into a streaked pile of concrete and brick dust. A youth stood there gazing at him, heel of one foot hooked up against what little of the wall remained. Hawkins took his folded paper from his pocket and offered it to the boy who looked insolently at it, then back up at him without any expression.

'Can you help me?' said Hawkins. 'Is this the street?'

When there was still no response Hawkins repeated the question in French, the only other language he had.

'It's here,' said the youth finally, in English.

'Where?' said Hawkins.

The youth twitched his hand along the road in which they stood.

'I am looking for Ninh,' said Hawkins. 'Nguyen van Ninh.'

There was another empty shrug and Hawkins wondered whether he should offer the youth money. For what? he thought. Instead he pushed through the empty gate, forcing the boy to move aside and went up the overgrown pathway towards the main house. Hutments and shelters were erected in the grounds, their owners squatting or standing around and when he reached the house Hawkins saw that it was a commune, occupied by several families. The house was skirted by an open verandah and people were assembled upon it in separate groups, each preserving their own territory. He went up the broad stairway and smiled hopefully about him. The answering reception varied. The children continued to giggle, as the others had done outside in the street and several of the women smiled shyly back at him. But from the men the response was reserve, although not as openly hostile as the youth at the gate. Hawkins moved cautiously among them, anxious against causing offence by intruding on to their patches, repeating the name of the man he wanted and receiving shrugs and head shakes in return.

The youth had gone when Hawkins left, and he felt a fleeting relief. He repeated the questioning at the two remaining villas, human beehives like the first, and at both met the same blankness. As a belated thought, at each, he left his name and the address of the Caravelle.

Back in the street he looked undecided at the rash of smaller huts and buildings. He must have questioned twenty different families – maybe more – and the empty sun was balanced redly just on the horizon, already canopied by the purple twilight that he knew would be a brief, almost imperceptible moment before the darkness.

There would be others like the hostile youth and he was

quite alone, Hawkins realised. If the people he'd already questioned didn't know Ninh then it was unlikely the others would either. Maybe he'd try again tomorrow, in the safety of daylight.

He went more hurriedly back along the river bank road. The night had come with the shutter-quick darkness of Asia. Lights were pricked out among the harbour moorings but insufficient to determine the shape of any ship. Hawkins walked attentively, listening for the sound of pursuing footsteps, relieved when he reached the bottom of Tu Do, with better street lighting.

He regained the Caravelle before the return of the tour, showering away the perspiration of the afternoon walk. He hadn't tried hard enough, he knew: not as hard as his father would have tried, in similar circumstances. He'd come knowing it was unlikely that he'd be able to find the man but three or four hours and a few people questioned was insufficient. He'd make another attempt tomorrow, he decided.

Hawkins was standing with just the towel around him when the telephone rang and he answered it expecting Mr Tho or even the persistent German girl who was still flirting over the bath plug.

Instead a man said, 'You are from Edward Hawkins?'

'Who is this?' he said.

'Did Edward Hawkins send you?'

'Is this Ninh?' A hopeful expectation moved through him. When there was no immediate reply Hawkins repeated, 'Is this Nguyen Ninh?'

'Yes,' said the man. There was a pause and then he said: 'I knew he wouldn't forget.'

Elliott Blair sat at the front of the droning transporter, looking back at the group he'd led through the two week exercise in the jungle of Panama, confident ahead of the final assessment that they'd done well.

It had been a survival test so they'd gone in with nothing, just the halazone tablets to purify the river or ditch water and their knives the only weapons, forced the trap or ensnared what they wanted to eat.

The pursuit group had been given their embarkation coordinates within an hour of their setting off but Blair's group had managed to evade for eight days and then it had been

Blair's men who sprang the ambush, the day before yesterday.

That hadn't been part of the original briefing and Blair knew the others would bitch about him being a smart-ass but he didn't give a damn. A proper Green Beret didn't play like it was some game, with a set of rules: innovation and manoeuvrability – hitting the other bastards before they hit you – was as much part of survival as spearing fish or snaring a rabbit with a fall-pole and a length of twine. More so, in fact. So it was right they should learn about it. That was the point of exercises, after all. Blair sighed, caught by the word. He was unhappy, at the limitation of exercises. It didn't matter how hard they were set – whether there was live ammunition or whether a man ate or starved upon the degree of his expertise – that's all they were, just exercises. And an exercise was never as good as the real thing.

Blair missed Vietnam. He missed being stretched, every hour of every day of every week, knowing that the slightest mistake, a blink when he shouldn't have moved, would have been the end. *That* was how to learn how to survive: what it meant to be a member of the Special Forces. Exercises could never properly fill in the gaps, teach a man how to react *instinctively* to a situation. In 'Nam a bunch of Green Berets wouldn't have been trapped as he'd trapped their pursuers, crouched around on their butts hauling on cigarettes and reminiscing about the last piece of tail they'd screwed. And the assessors would recognise it, in the final reports.

Blair looked away from the men of his immediate command, trying to define how he felt about them, unhappy at the word when it came to him. Pity was the one that fitted, he decided. He pitied them for not knowing what 'Nam was like; pitied them like hell.

Chapter Eleven

Hawkins maintained the pretence of a stomach upset to avoid the delta tour, remaining at the window of his room until he saw the coaches pull away below. Heat was already beginning to fill the day when he got outside the hotel and he walked unhurriedly, well ahead of the time Ninh stipulated. He crossed Tu Do and cut through the linking alley to get to the rue Catinet. He turned immediately left, going back towards the river, excited at having located the man. At once came a balancing caution. He *hadn't* located Ninh: not properly. The man still had to keep the appointment. Ninh had been clearly nervous on the telephone, refusing to say where he lived and insisting upon practically theatrical meeting instructions.

At the bottom of the broad avenue Hawkins turned obediently right, parallel with the river again. It was busy with bustling boats and the quays and jetties swarmed with people. Ahead he picked out several fish stalls and realised he was entering the market area that Ninh had talked of in their stilted conversation the previous night. Following the instructions he went through it and then paused on the outer edge, staring into the yellow water. It was full of refuse and stank of sewage again. The crane Ninh had provided as another marker was about twenty yards away, oddly painted in some camouflaged way, a dated relic of the war. Hawkins supposed it would have originally been American equipment: he'd never thought of cranes as being things protected by camouflage. He went to it, grateful for the narrow strip of shade that came from a grated platform about thirty feet from its mobile tracking. There was a blast from the river and Hawkins looked towards the sound, watching one of the East German freighters start to make way down the river. Hawkins checked his watch and saw he was fifteen minutes ahead of the meeting time. What would he do if Ninh didn't keep the meeting? Go back to the shanty area, he supposed. Which of the families of the previous day had known Ninh and passed on the message?

A stooped, elderly man approached, seeming to need the support of the bicycle he was pushing. It was a rusting, rattling machine, without a cross-bar and with most of the covering split away from the seat, exposing metal ribs. There were some vegetables tied to the handlebars. The man propped the bicycle against one of the crane legs and fumbled with a pump against the rear wheel: Hawkins hoped he wouldn't remain there too long and frighten away the Vietnamese he was waiting to meet.

'I am Ninh.'

He spoke still bending over the bicycle, not looking up and Hawkins physically jumped, startled. He squinted against the outside glare of the day, focussing on the frail figure. Only two years between them, Hawkins remembered, from the documentation in his father's trunk: from Ninh's appearance Hawkins thought the difference could have been twenty. Stick-thin arms protruded from a stained, fraying shirt and his shoulders were bowed and hunched, as if he'd spent a lifetime carrying heavy burdens. There was a shank of pure white hair running back from his left temple.

'Ninh!'

The man came up to him at last, looking apprehensively along the jetty to ensure they were not being observed. He smiled, a gap-toothed expression and said 'You look like him. He told me about you.'

Hawkins was confused by Ninh's appearance: having achieved the encounter he didn't know how to pursue it. He said, 'Is it dangerous for you, our meeting?'

'Yes,' said Ninh, simply. 'You asked many people about me yesterday. Too many.'

'I'm sorry if I've caused you trouble,' said Hawkins.

Ninh humped his bony shoulders, to indicate the problem could not be undone. 'There are a lot of informers,' he said. 'One is taught that loyalty is proven through vigilance.'

The man spoke as if he were reciting a lesson, thought Hawkins. He said, 'You have been unwell?'

Ninh nodded, 'Malaria; I am better now.' In immediate

contradiction Ninh was seized by a sudden rasp of coughing, a hollow-chested, consumptive sound.

Hawkins said, 'Are you having treatment?'

'It is a cold, nothing more,' dismissed Ninh. Then he said anxiously, 'What does your father say: what is the arrangement?'

Hawkins crossed the shaded area of the crane, coming close to Ninh and instinctively the Vietnamese once more checked the jetty on either side of them. 'I'm sorry,' said Hawkins. 'I don't understand.'

The other man put his head to one side, looking at Hawkins curiously. 'There must be an arrangement,' he said, the plea edging into his voice. 'Your father promised to help me: act as guarantor, for me to leave here. I've waited so long; so very long.'

Hawkins swallowed, not knowing how to respond. Surely Ninh had to be mistaken, to have misunderstood! There had never been any conversation with his father about bringing the interpreter out of the country. And it was inconceivable that his father would not have talked about it, if such a promise had been made. Ninh *had* to be mistaken.

'My father's dead,' said Hawkins.

'Dead?' The man shook his head in disbelief, his face twisted as if he were in physical pain. 'Not dead,' he mumbled. 'Promised to help.'

'I'm sorry,' said Hawkins, inadequately.

'He was trying?' persisted Ninh hopefully. 'He was trying before he died?'

Imagining that he was meeting the man's need, Hawkins stupidly said, 'Yes, I think he was trying.'

A smile illuminated Ninh's face. 'Then it's still possible: you are going to help me?'

Could he? wondered Hawkins. He said, 'It's not easy. Very difficult, in fact.' What about Peterson? 'There is someone, an important man; I will speak to him.'

Was he being cruel, giving such an undertaking?

It was something he could do, easily enough. But he couldn't anticipate that Peterson would help; that he could help, even. And his own situation created a complication. He was an Englishman, on temporary assignment in America. Would the US immigration authorities accept his being a sponsor to

Vietnamese? Unlikely Hawkins decided, objectively. So what could he do?

'Thank you,' said Ninh fervently. 'You are a good man, like your father was a good man.'

'How has it been, since the end of the war?'

Ninh's face twisted again. 'Not easy,' he said. 'They said I had to be re-educated: that I was guilty of errors . . .'

'You were imprisoned?' cut off Hawkins.

'Re-education camp,' insisted Ninh. 'Then the delta. Pleiki in the Central Highlands. I contracted the malaria there.'

And became prematurely aged, thought Hawkins. He said, 'Were you sent to the camp because of your association with my father?'

'It was not his fault,' said Ninh defensively. 'He was not to know: none of us were. That's why I have to be careful now.'

Hawkins recognised that he had some sort of moral obligation to this man: he'd definitely try to get Peterson's assistance. Wanting to do something positive, Hawkins took all the American money he possessed, about $120, from his pocket and said, 'I want you to have this. I'm sorry it's not more. Can you convert it, like before?'

He knew that during the American presence there had been war script as well as the Vietnamese piastre but that genuine American dollars could be exchanged for four or five times its face value on the black market. Ninh looked reluctantly at the proffered money and Hawkins realised that the man was trying to retain some sort of dignity about the charity. Ninh snatched out almost angrily, taking it and thrusting it at once into his pocket. He said 'It can still be changed.'

'Perhaps I can send you more, when I get back?' said Hawkins. As an afterthought he said 'I live in Washington.'

'You should be able to help me, if you live there,' pounced Ninh.

'I said I'll try,' repeated the Englishman. 'I'll need your address.'

'It wouldn't be safe, for you to write,' warned Ninh. 'The letter would be intercepted: I could be accused of spying.'

'How can I contact you then?'

'Only if there's news about my leaving.'

'All right,' said Hawkins, uncomfortably. 'How?'

'A letter then: the street number is fifteen.'

The last house he had visited the previous evening, Hawkins remembered. He said, 'You were there, when I came?'

Ninh nodded, lowering his head.

'Why did people say they didn't know you? Why didn't you come forward?'

There was another uncertain shoulder movement. Quietly Ninh said, 'Everyone there has been to a camp, at sometime or another. No one knew if it was safe; if I wanted you to find me . . .' Ninh hesitated and then went on 'And I was ashamed of where I am. It is not like before, when your father was here. I had my own apartment then.'

'I tried for my father's old room at the Caravelle but they said it wasn't possible,' disclosed Hawkins.

There was another look of curiosity from Ninh. 'Your father only stayed at the Caravelle in the early weeks,' he said. 'After he met Nicole they took one of the villas near the Cercle Sportif: they used to like the club.'

Hawkins gazed at the other man for several moments. Finally he said, 'Nicole?'

'Nicole Tiné,' said Ninh patiently. 'Your father was with her all the time, right up until the end.'

'Where?' groped Hawkins, his mind reeling with uncertainty. 'Where is the woman now?'

Up and down went the man's shoulders. 'I suppose your father tried to get her out: I know he intended to because we talked about it, the three of us, just before the end. Now I don't know where she is.'

'You mean that somewhere here, in this city, is a woman who was my father's wife!'

'Never married,' said Ninh dogmatically. 'Maybe she dead, maybe she alive. I don't know.'

His father had been a widower by the time of his second period in Vietnam so Hawkins felt no moral surprise at learning about a woman named Nicole Tiné; and morality didn't really enter into it anyway. But he felt outraged; cheated, too. Feeling cheated was the worst. He'd lived with his father for the last two years of the man's life; been open and honest with him about everything and believed his father had been open and honest with him in return. And now he knew the relationship he had imagined had been just that, imagined. What had existed between them was nothing more than a surface acquaintanceship of casual friends. *Why!* he thought angrily. Why the hell hadn't the man told him about a woman with whom he'd lived, as husband and wife, for two years? And about this crushed, shuffling man with the constantly nervous eyes whom he'd promised to get out to the West. *Had* his father promised? Hawkins realised there was only Ninh's word: just as he only had Ninh's word for the existence of someone called Nicole Tiné. But what advantage was there for the man lying, about either? There was no benefit: danger, in fact, in making the meeting in the first place, if Ninh were to be believed about the re-education camps. He'd even appeared reluctant to accept the small change Hawkins carried in his pocket. Hawkins recognised that he was confused: confused and uncertain.

It was very hot now, even in the shade of the crane. He shifted uncomfortably and said to the Vietnamese, 'Isn't there somewhere else we can go? I need to talk to you.'

Ninh shook his head. 'Not safe,' he insisted.

'Tell me about the woman,' said Hawkins.

'Pretty girl,' said Ninh, smiling in reminiscence. 'Father was with the French legion, in 1950s, Mother was Vietnamese. They had a pretty house in Dalat but Nicole worked here, at the French embassy. That's how your father met her, at a reception.'

Hawkins frowned and said, 'If her father was here with the French in 1950 she must have been young.'

'Twenty, when they met,' agreed Ninh.

So his father had been twice as old. He said, 'They lived near the Cercle Sportif?'

'The villa with green shutters, almost directly opposite. They were very happy there: there were parties. Good times.'

Caught by a sudden thought, Hawkins said, 'Do you have a photograph?'

'No more,' said the man. 'Maybe once but no more; not safe to have such things.'

There was still the Vietnam trunk to go through completely, remembered Hawkins. Would he find a picture filed there? 'You don't know the camp to which she was sent?'

'We've talked too long: I must go,' said Ninh.

'No!' said Hawkins anxiously. 'Not yet. Wait a little longer, please.'

'It's dangerous.'

'Are you *sure* she was sent to a camp?'

'That's what I heard,' said the Vietnamese. 'When the Americans left everyone was very frightened; didn't know what to expect. Nicole went to Dalat, to be with her mother. I heard she was arrested there.' He made an expansive gesture. 'People got sent all over: Hanoi even.'

'If her father was French she would have had French nationality,' said Hawkins, with sudden awareness. 'Why didn't she get out?'

'Her mother didn't want to go.'

'I want to find her,' Hawkins said.

'It would cause her harm,' insisted Ninh.

'If I try to help you . . . and I promise that I will . . . can you try to find out about her. To see if there is anything I can do.'

'I will try,' said Ninh but Hawkins wasn't convinced the man would.

'Your word?' demanded Hawkins.

'We were once friends,' said Ninh. 'I will try.'

'They were happy?'

'At the villa, yes. Always. Only the baby was sadness.'

'Baby!'

'Why do you not know this?'

'My father and I were not often together: I was away,' lied Hawkins.

'A little girl, Elian.'

'What happened?'

'She only lived a month . . .' Ninh held himself and said 'Something wrong with the stomach, here. Your father got her into the American hospital but she died. Then they were sad.'

'Did you know about me?' demanded Hawkins abruptly. 'Did my father speak of having a son.'

'He talked about you,' said Ninh.

'Tell me about how he worked: what he did,' asked Hawkins.

Ninh shook his head, positively. 'No more time: we've spent too long here.'

'Tomorrow,' pleaded Hawkins. 'Meet me tomorrow; I've only got two more days in Saigon.'

'Ho Chi Minh City,' said Ninh. 'It is necessary to be correct.'

'Only two more days here,' repeated Hawkins.

'This place, the same time tomorrow,' agreed Ninh. 'But you must not come to the house. It is not safe.'

'I won't come to the house,' promised Hawkins.

'You'll try to help me!'

'Yes,' said Hawkins solemnly. 'I'll try to help you.'

Back at the hotel Hawkins sat in the gradually darkening room, trying to absorb what he had learned. Cheated, he thought again: utterly cheated. He possessed cases of material from which he thought he knew everything there was to know about his father and he realised he knew him not at all. How many other Nicole Tinés and Nguyen Ninhs had there been, in Korea and Algeria and Africa and the Mediterranean? Had there been other babies, babies who could have survived, to be his half-brother or half-sister?

He left the hotel early the following morning, planning another visit in advance of the harbour meeting. The Cercle Sportif had changed from the social hub it had been during the French and American occupation, appearing now to be a meeting place for political gatherings. Hawkins ignored it, disinterested, looking instead for the villa with the green shutters.

It was where Ninh had described it, fading in the sunshine like most of the houses appeared to be doing. It was a large building; from the street Hawkins guessed it could have as many as six or maybe eight bedrooms, so below there would be several reception rooms.

His father had lived there, thought Hawkins: lived there and loved there and made a secret life there. *Why* hadn't his father told him!

He considered going in but decided against it. He didn't have the language if the Vietnamese there didn't speak English or French and if the woman had survived he could be endangering her, by making enquiries.

Hawkins took several photographs and left with more than sufficient time to get to the harbour. He approached the mottle-painted crane familiarly, settling gratefully into the shade. He wanted to know so much. He wanted to know about his father's work-day, his normal routine: how often he travelled around the country and if he ever took Nicole with him: whether as far as Ninh was aware, there had ever been discussion of marriage between his father and the woman: how he'd relaxed, spent his spare time; if he'd been liked, by the other correspondents; the sort of man he'd appeared to be, with other people; whether he'd drunk too much or smoked too much; everything, thought Hawkins, everything there was to know. He'd tried, as best he could, to erase from his mind the impressions and the recollections, leaving himself with just the outline shape of a man. He wanted Ninh to colour it in, provide the features and the attitudes.

The previous day Ninh had been on time but today he wasn't. At first Hawkins was not overly concerned: there were a dozen reasons why the man should have been delayed. After thirty minutes, he began to stir, staring up and down the marketplace harbour, anxious to see the emaciated man with the bicycle. After an hour Hawkins felt a burn of annoyance, a physical impression beyond the discomfort of the overhead sun. Damn the man, for not coming! He had so little and wanted so much more. There was always the address: number fifteen, the last crumbling villa in the line.

And then he remembered the unremitting nervousness, the way the eyes had pebbled at the first sign of movement from the crowded area further along the quay. With one recollection came another. Four years in a re-education camp, because of his association with a Westerner. '*One is taught that loyalty is proven through vigilance*' Ninh had said. Was he prepared to risk subjecting the man to another four years of re-education, another sentence to age and break him further?

No, he thought. He wasn't prepared to do it. Even though the supposedly easy book wasn't easy any more and even though he knew he should as a professional reporter, he wasn't prepared to do it.

That night, for the first time for weeks, Ray Hawkins got drunk, a maudlin, mechanical process, alone in his room. Before it got too bad, while there was still some coherence in his thoughts, he decided it had been a mistake to come. A bad mistake. And that he didn't know what to do about it.

'November,' announced Sharon Bartel.

'That's six months away,' protested Patton.

'You agreed I could decide when,' reminded the woman.

'What about the apartment?' he said. It was on Riverside, on the twentieth floor, with a spectacular view of the river.

'It's too expensive,' she said.

'No, it's not.'

'It's four hundred thousand, for Christ's sake!'

'Do you want it?'

She made an indecisive gesture, not looking at him. 'I'm frightened, spending that much money.'

'I can afford it . . .' He shook his head, in correction. '*We* can afford it.'

She felt out for his hand. She said, 'I'm so nervous about moving on, from what we've got already. Of taking chances. So everything gets confused because spending money is taking chances, even if you've got it to spend.'

'You're not making sense,' he said, softly.

'I know I'm not.'

'But I know what you mean.'

'Do you?' she said pleadingly, coming back to him.

'Yes,' he said. 'You won't lose out again, I promise you. From now on it's going to be happiness, all the way.'

'I do love you,' she said.

'I'm going to buy the apartment,' he said.

'I'm looking forward to living in it.'

Chapter Twelve

As he had earlier, seeking the video recording of the orphan rescue and the notebook in which that recording showed his father to be writing, Hawkins did not confine himself to one case. He made a detailed, document-by-document search through all his father's effects: he'd been careful, about everything, but it was possible he'd overlooked something, that there could be a paper or a file or some official notation tucked inside another folder or accidently hidden in a ledger. Anything, in fact, to make him feel he hadn't been cheated. But there was nothing so he started feeling cheated: nothing about Nicole Tiné and nothing to indicate his father tried to get Nguyen van Ninh out of Vietnam. Nothing. About the woman Hawkins accepted he was helpless.But about Ninh he could at least try to keep the instinctive, conscience-spurred promise of the Saigon dockside.

John Peterson was the key, but Hawkins was unsure how to approach the senator. He supposed the assistance he wanted was official yet the inclination was to make the request unofficially, through the Georgetown number rather than the Senate office. He was still unresolved when the uncertainty was settled for him by the unexpected telephone call from Eleanor Peterson.

Throughout the trip to Vietnam Hawkins had consciously kept the woman from his mind, not wanting to be reminded of an embarrassment. Which was how it had to be regarded. It had been an embarrassing mistake to invite her to eat with him and an embarrassing, reckless mistake to have kissed her. But it had only been a kiss. So that was all it remained, an embarrassment. Was Eleanor Peterson a shy, genuinely lonely woman who'd momentarily sought friendship the wrong way? Or a spoiled, rich neurotic playing soap opera games. He didn't think she was playing but he wasn't sure. And he was determined – positively, adamantly determined this time – not to try to find out. Things were professionally going too well to risk it all for some meaningless affair.

It was good to hear her voice.

'How was the trip?'

'Pretty good,' avoided Hawkins. She'd have to know sometime, he supposed, but he didn't feel like telling her on the telephone. He wondered what she was wearing.

'I came over several times, while you were away,' she said.

Giving her the key to Maryland Avenue while he was in Vietnam had been part of the resolve not to become involved. 'All through?' he said.

'I guess so.'

'Good,' he said, not able to think of anything else.

'I'd like to come over,' she blurted.

No! he thought. Despite all the resolution he wanted her to and because he wanted her to he knew he should refuse, should say 'No' and mean it, not just think it. 'Sure: any time,' he said, trying to get the vagueness into his voice.

'I meant now.'

I know you meant now, he thought. He said: 'I'm writing a feature about the trip.' He'd decided against telling London about Ninh and certainly against including it in what he was writing for the newspaper: publicity at this stage might endanger his chances of getting the man out. 'And I really should get on with the book,' he added.

'There's something I want to talk to you about.'

'Problem?'

'To me it is,' she said.

It was childish, physically trying to distance himself from her. He knew the dangers so all he had to do was avoid them. Easy enough, with the newly found self-control. 'OK,' he said.

'Thirty minutes?'

'Thirty minutes,' he agreed.

She was five minutes late and he was impatient. She was bundled in a quilted topcoat with the familiar woollen hat tugged down over her ears and when she took it off she merely shook her hair loose, running her fingers through it. Hawkins saw she wasn't wearing any make-up and wondered if the redness of her eyes was just because of the cold.

Without prompting she went into the jumbled study and sat demurely on the couch where last time they'd kissed. He thought she looked beautiful and wished he didn't.

'Yes,' she said, in advance of his suggesting it, 'I could do with a drink.'

'Upset?'

'Yes.'

'Why?'

'You know what I think about politics?' she asked rhetorically. 'I think politics is an asshole.'

Hawkins gave her the Scotch and said 'What's happened?'

'John's broken his promise.'

'What promise?' frowned Hawkins.

'About John junior,' she said. 'About not using him in the campaign. He's agreed to a photo-session for *Women's Wear Daily* . . . "The Contender at Home with his Family", that sort of thing.'

Hawkins grimaced and said, 'Why?'

'Falling poll ratings. Can you imagine that! He's doing it because of falling poll ratings. Jesus!'

'Haven't you tried to dissuade him?'

'Of course I have,' she said impatiently. 'He says there's no harm: that it won't hurt.'

'Maybe it won't, just once,' said Hawkins.

'But it won't be once, will it? Whenever Joe Rampallie or Pete Elliston or anyone else wants some fucking gimmick, out we'll all be trotted again, to perform.'

He'd been very wrong about her swearing, he decided. He said, 'Maybe it won't be like that,' realising it was inadequate.

'He *promised*!' said Eleanor vehemently. 'He knows how I feel and he agreed to keep the family out of it. Certainly John junior. I think it stinks. It all stinks.'

She'd put him into a position of confidant again, Hawkins realised. He put his hand out for hers, a consoling gesture, but she took it and held it tightly, and he was excited by the touch. 'I want John's help,' he said, intruding her husband between them.

'How?' she said.

It took him a long time. He began badly, intending only to

tell her about Ninh but then the account became awkward and he decided he needed to talk it through with someone, like she had talked through her problem with him and so he told her everything. She listened without interruption, gazing down at his hand, tracing her fingers over his. When he finished she said, 'That's amazing. Absolutely amazing.'

'Yes,' agreed Hawkins bitterly. 'Bloody amazing.'

'How do you feel?' she asked. 'About his having another woman. A baby?'

'I don't have any feeling about that,' he said. 'Not in the way you mean. It wasn't like that, between us: couldn't be. I just can't understand why he never told me . . . why he never tried to help them.'

'You don't know that he didn't,' said Eleanor. 'All you know is that there isn't any record, in all the stuff in the boxes downstairs. Maybe he did try. But for some reason didn't keep a record.'

'You've seen the trunks: read what's in them,' said Hawkins, refusing her argument. 'He's got laundry bills going back fifteen years, for Christ's sake! If he's going to keep something as inconsequential as some of the stuff down there don't you think he'd have kept a file on his efforts to get out a woman who bore his child. And a man to whom he'd made a promise to take to safety.'

'We seem to be talking a lot about broken promises, don't we?' said Eleanor, reflectively. 'And yes, I would have expected him to keep files.'

'Why *didn't* he!' demanded Hawkins, exasperated. 'I was with him for two years, for God's sake! Not once in two years did he as much as mention another wife or another baby. That's not natural.'

'What if he *had* tried? And failed. What if he knew he'd lost them?' suggested the woman. 'Perhaps he was frightened that by telling you he'd lose you too.'

'He didn't have any reason for thinking like that,' said Hawkins. 'I told you, it wasn't like that between us.'

'Not on your side,' she said. 'What about his? Do you know how he felt?'

Hawkins gave a dismissive laugh. 'I'm damned sure I don't know how he felt. I thought I did but now I know I didn't have a clue.'

'I wonder what happened to her,' said Eleanor, distantly. She shuddered. 'Poor woman.'

'I wish to God I knew,' said Hawkins. 'I don't even know if I've caused Ninh more trouble, by making contact. Do you think John can help?'

She shrugged. 'Probably. You can ask him.' Bitterly she added, 'There might even be some election advantage in it.'

He was enjoying her closeness too much. He said, 'If it gets Ninh out of Vietnam I wouldn't see any harm in that.'

'John wants you to come to dinner sometime,' she announced. 'Suggested I call you.'

'I'd like that,' said Hawkins. He looked at his watch and said, 'It's getting late.' He wanted her to go, before it *was* too late.

'John's in Denver,' she said. 'I saw the children before I left.'

'I could do steaks again.'

'I'm not really hungry. Are you?'

'Not really,' he said. 'Another drink?'

'Not particularly.'

There was a loud silence between them. Eleanor broke it. She said, 'I'm sure,' and then she lifted his hand to her lips and kissed him, softly.

'I decided this wasn't going to happen,' he said. 'I don't want it to happen.'

'Is it?' she said.

'Yes,' he said.

There was no anxiousness between them. They came together quite relaxed, exploring and playing and enjoying, each wanting to please the other and when they did they knew they could do it again and so they did, more experimentally this time. Hawkins was surprised at Eleanor's sexuality and of his ability to match it: pleased, too.

Afterwards he said, breathlessly, 'That was wonderful.'

'For me, too,' she said.

'How's it going to be, when I come to your house to meet John?'

He felt her shrug beside him. 'I don't know,' she said. 'Will you be embarrassed?'

'I don't know either,' said Hawkins.

'John isn't interested any more. He's got a travelling harem of helpers.'

He wedged himself on his elbow, so he could look down at her. He kissed her, nipping, exploring kisses and said, 'How the hell is this going to work out?'

'I'm not going to worry about it working out,' she said. 'I'm just going to enjoy it, from day to day.' She was quiet for several moments and then she said, 'I meant this to happen. That's why I said I wanted to read your father's stuff and made it so you had to invite me. I didn't really want to read it at all. Not much anyway.'

Hawkins wasn't sure he would ever become reconciled to her honesty. 'Why?' he said.

'You're not strong, like I'm not strong,' she said. 'You try hard like I try hard and if you succeed you give yourself a reward which is just a convoluted way of being weak again. Which isn't anything to be ashamed about. Or embarrassed. Or to deny. I'm fed up to here with supermen who know how every day's going to end at the very moment it begins: I'm fed up with control and reserve and moderation. And worrying, every time I open my mouth, whether I'm going to get it right or whether I'm going to see a frown of disapproval . . .' She raised herself from the bed, to kiss him. 'I think you're ordinary and human and get it wrong as often as you get it right and I don't think you sit in judgment upon me . . . not much, anyway . . . and I enjoy being with you and now it's happened I also like screwing with you.' She kissed him again and said, 'I'm not offering myself as an expert but I thought the screwing was terrific.' She fell back beneath him against the bed and said, 'And for me that was a hell of a speech and I hope you're not offended.'

Now Hawkins kissed her. He said, 'No, I'm not offended. Although I think I should be. You've just admitted to using me, absolutely.'

'Yes,' she said, 'I have.'

'Use me any time,' said Hawkins. 'All the time.' Despite which he wished the weaknesses weren't quite so evident.

Hawkins went to dinner at Georgetown the following week. Peterson greeted him effusively and Eleanor shook his hand with polite distance and Hawkins wondered why he wasn't feeling the embarrassment that he should be feeling. It was a large party, eleven more guests apart from himself. He was paired with a brightly attractive, intense campaign worker for Peterson. Her name was Beth and she flattered Hawkins by obviously knowing of his work. She gave him her telephone number when they were at the dessert stage, before the women followed the English practice of leaving the room for the men to smoke cigars and drink port or brandy. Deciding it was the most opportune moment, Hawkins made his approach to the senator, abbreviating the story this time only to include Nguyen van Ninh.

Peterson listened with head-bent attentiveness, rolling a thin cheroot-type cigar between his fingers. He said, 'That isn't going to be an easy one, Ray.'

'I know that,' said Hawkins. 'I'm anxious to do whatever I can.'

'I understand that,' said Peterson. 'Leave it with me. I won't make any promises but I'll see what I can do.'

'I appreciate it,' said Hawkins.

Paul Brunton, the Speaker of the House who sought Hawkins out after his interview with the President, was one of the three other Congressmen at the dinner in addition to Peterson and even when they rejoined the women the conversation remained political. The women seemed to expect it and Hawkins welcomed it, regarding it as exactly the sort of background contact necessary for him to fulfil his function in Washington. Towards the end of the evening, he found himself temporarily alone with Brunton, near the book-clustered shelves of the drawing room.

'Hear there's a problem with the President's inspection plan,' offered the politician.

Brunton was a Peterson supporter and Hawkins wondered if he were being purposely fed: they seemed to be conveniently alone in a crowded room. He said, 'What sort of problem?'

Brunton made a vague, uncertain gesture. 'Gather your people and other Europeans aren't happy with the Soviet response.'

'Why not?' said Hawkins.

'Only corridor gossip, you understand?'

'I understand.'

'Word has it that the Russians have asked for site inspection in Britain, France and Germany and that they've all said no: complained it's overlapping sovereignty,' said Brunton.

He *was* being fed, Hawkins decided: well fed. 'Thank you for telling me,' he said.

'Didn't come from me, you understand,' said Brunton hurriedly.

'Of course not,' accepted Hawkins.

'Be quite a setback for the plans, wouldn't it?' said Brunton, moving away to join another group near the patio windows.

When it came time to go, Beth, who had remained comparatively close to him throughout the evening, apologised for having to drive home two of the other campaign people who had accompanied her to the party and Hawkins said he had his own car anyway. She said she hoped they'd meet again real soon and Hawkins said he hoped so too, real soon.

At his own moment of parting Hawkins was alone with Eleanor for the first time during the evening. She looked at him with her head to one side and said, 'If you use that telephone number she gave you I'll kill you.'

He laughed back at her and said, 'That was supposed to be discreet.'

'She was practically on her back for you,' said Eleanor. 'And I was as jealous as hell.'

'I can't even remember what the number was,' said Hawkins.

'Liar!' she said.

'I love you,' he said.

The smile went from her face and she became suddenly serious. 'I wish you hadn't said that,' she told him.

Chapter Thirteen

On his way along Massachusetts Avenue towards the great mansion of the British Embassy Hawkins considered asking the ambassador for help in getting Nguyen van Ninh out of Vietnam. There would have to be some contact between them if he, as a Briton, were going to act as a sponsor in America. He decided against it. Peterson had asked that it should be left to him and Hawkins recognised the risk of confusion if he involved too many people too early. The time to discuss it with Sir Neville Wilkinson would be after he learned what Peterson could do.

Hawkins was ushered straight in to see the ambassador, in a study of ornately panelled and carved wood, oak or mahogany he guessed, overlooking the Naval Observatory which forms the official residence of the American Vice President. Coffee was already set out and Jeremy Abbott poured. There were three cups and Hawkins realised the press counsellor intended staying, just as John Folger had remained for his meeting with the President, to act as a witness.

'I'm sorry I wasn't able to make the other meeting,' said Hawkins. An invitation to one of Wilkinson's briefings had been among his mail when he returned from Vietnam: he'd already apologised by letter.

'How was the trip?' said the ambassador.

'Interesting,' said Hawkins. 'I would have liked to have had personal knowledge of what it was like before, to ensure my comparison was accurate.' And not felt so hindered by the knowledge of Nguyen van Ninh and Nicole Tiné, he thought.

'Are you going to write about it?'

'It's due to appear on Sunday,' said Hawkins. There had

already been what now seemed to be the predictable congratulatory cables from London about the breadth of the piece but Hawkins was unhappy with it. He felt he had written a bland, holiday supplement type of feature.

'I'll be interested to read it,' said the ambassador, politely. Indicating the press counsellor, Wilkinson said, 'I gather from Jeremy there's something specific you want to raise with me?'

One of his father's earliest lessons had been that people always offered most if they believed the questioner had more information than he really did. Hawkins said, 'I'm writing a lengthy piece for Sunday on the collapse of President Harriman's peace initiative, because of the European difficulty over inspection. So I naturally wanted to make sure I got the British perspective completely right.'

Wilkinson frowned across the room at his press counsellor and then said to Hawkins, 'You seem to be most remarkably well informed.'

'Thank you,' said Hawkins and stopped, refusing to be flattered away from the question: silences produced more than quotations, his father often said.

'I would consider it an exaggeration to say that the missile inspection proposal has *collapsed*,' said Wilkinson, filling the gap.

'Why?' said Hawkins, keeping the ambassador in the position of having to offer information.

'Because it is my understanding that discussions and negotiations are continuing,' said Wilkinson. 'Any negotiation of this magnitude is bound to encounter stumbling blocks.'

'Is that how you regard what's happened with Britain – and the rest of Europe – just a stumbling block?' said Hawkins. It was a weak question, he knew, and one that Wilkinson could avoid if he recognised it.

'Great Britain has quite rightly told the administration here that under no circumstances could it consider negotiations in Geneva – negotiations in which it has not got a direct, contributory part – resolving the right for Russian inspection upon its territory. And I think that is an attitude that the administration understand: the United States would not, after all, allow Britain to discuss with a third party access to America,' said Wilkinson.

He'd got away with it! thought Hawkins: the ambassador

hadn't realised the weakness and given him the confirmation. Exaggerating his knowledge further Hawkins said, 'But if Britain – and the rest of Europe – are refusing the Russian insistence for on-site inspection in their countries, then how *can* the negotiations proceed? That's not a stumbling block: it's a stalemate.'

'There's always the American compromise,' said Wilkinson.

Damn, thought Hawkins. Searching, he said, 'I'm not convinced the compromise will solve everything.'

'If there aren't any American missiles in Europe then there are no sites to inspect, are there?' said Wilkinson. 'How can you fail to be convinced by that?'

Wilkinson had missed it a second time, Hawkins thought. A few weeks ago Hawkins might have allowed his euphoria to interrupt his questioning but not now. He said, 'America can take its missiles out of the NATO countries of Europe but France isn't a member of NATO and it's got its own, independent deterrent. So has Britain. And does Britain *want* to be stripped of US missiles, as a concession to Russia?'

Wilkinson nodded in admiration and said, 'You're a persuasive debater, Mr Hawkins.'

Again Hawkins refused to be flattered. 'I still see it as a stalemate,' he said.

'I agree there's a great deal more detailed discussion to be conducted in Switzerland before the problems can be resolved,' said Wilkinson. 'I think it's sad. I think the President's plan was a bold one and deserved more success.'

It was John Folger's suggestion that he should meet Hawkins in the dining room of the Hay Adams Hotel, directly across Lafayette Park from the White House. Hawkins arrived first, to ensure the table he'd reserved by telephone was as secluded as he wanted. He didn't consider it was, so he changed to another directly in the corner of the panelled dining area, far enough away from any adjoining table for them not to be overheard.

Folger was early, too, thrusting into the dining room and staring around enquiringly, a slim, neat man in button-down shirt and three piece, Ivy League suit. Like so many political aides, Folger was the sort of man who would always need to hurry, to appear busy, thought Hawkins, as the presidential Press Secretary located him and came striding across the room.

There was the same effusive handshake as on the day of the interview with Harriman and a refusal to drink anything other than club soda. Hawkins kept the waiter, wanting to get the order out of the way, but went patiently through the conversational pleasantries with the American. Hawkins waited until the main course was served and then said, 'I'm going to write next weekend about the missile difficulties. About the Russian insistence upon European inspection and the President's offer to withdraw US missiles from NATO countries.'

Folger stopped with his fork suspended in front of his face. He said, 'Holy shit!'

'Is the administration demanding equal inspection rights, in the Warsaw Pact countries?' said Hawkins.

'Now pause here, Ray,' said the Press Secretary, raising his hand in a halting gesture. 'I think we've got a crisis.'

'That's what I'm writing about,' said Hawkins.

'I didn't mean that. I meant your knowledge of it: even *I'm* not sure of the full facts. You're going to make a lot of important people unhappy if you go into print with this.'

'Come on!' protested Hawkins. 'You can't expect me not to file something like this.'

'Wait at least for a balanced view,' pleaded Folger.

'That's why I invited you to lunch,' said Hawkins, easily.

'There's a problem, I'll concede that,' said Folger. Anxiously he added, 'Not officially, of course. But problems arise all the time in something as complicated as this: I think you'll be making a professional mistake to let go both barrels at the same time.'

'That's a cheap shot, and you know it,' said Hawkins forcefully. 'I'm a working journalist with a story. Either I go into print with your side or I go into print without it on what I know, which is considerable . . .' He paused and said pointedly, 'Which isn't the sort of blackmail you tried a few moments ago but me speaking honestly as the working journalist I've just told you I am.'

'OK,' said Folger, replacing his fork on to the plate of an unwanted meal. 'I shot my mouth off a little too soon. I made a mistake.'

'You've got a problem,' said Hawkins. 'You went public with the offer and now it's going to hang around your neck, like an albatross.'

'Negotiations are *continuing*,' insisted the press spokesman. 'Nothing's broken down.'

'You've got Europe split every way but which,' said Hawkins. 'The NATO countries relied upon the US and went through considerable political resistance installing the missiles you're now negotiating to take away. And that leaves them with their pants down. And you've got France able to thumb its nose, like France always thumbs its nose at everyone.'

'We've got State people in all the capitals, explaining everything,' said Folger, in an unknowing admission. 'Sometimes it's necessary to give initially to gain a lot later.'

'You getting permission to inspect in the Warsaw Pact countries?' said Hawkins, repeating his first question.

'Like I said,' insisted Folger. 'Sometimes it's better to give now to gain more later: we're sure we'll get the concession.'

'The peace protestors in Europe always claimed that America put its missiles there to use Europe as a buffer between itself and the Soviet Union; that Europe would be the first sacrifice,' said Hawkins. 'You're proving them right.'

'The governments will understand!' said Folger weakly.

'Western governments are elected by people,' said Hawkins. 'Will the people understand?'

'If it's explained properly to them.'

'So explain it to me.'

'The Soviets want a gesture,' said Folger. 'An indication of goodwill to show what we're offering is not just political polemic. We see their demand to inspect Britain and the other NATO countries as a feint, a poker-play. So we've responded to it. We've called their bluff by offering to withdraw our missiles in Europe.'

'With what guarantee?' pressed Hawkins.

'We expect a gesture from them.'

'That's not poker,' said Hawkins. 'That's a kid's version of Beggar-thy-Neighbour. Literally.'

'Not everything is being withdrawn,' said Folger. 'Just some.'

'Which country first?'

Folger shifted uncomfortably in his seat. 'Britain is regarded as a closest ally: always has been.'

He had the best so far, Hawkins realised. 'Have you got British agreement?'

'We expect to get it,' admitted Folger.

'You said you didn't know all the details,' persisted Hawkins. He realised he liked being so completely in control. And sober.

There was another uncomfortable movement from the man. 'That's it, broadly,' he said. 'Britain, then Germany, then Belgium, then a pause for some reciprocity.'

'What about France?'

'We're seeking cooperation.'

'You haven't finished your meal.'

'I'm not hungry.'

'I didn't mean to put you on a spot like this,' said Hawkins.

'I know,' said Folger, confronting a familiar explanation. 'You're just doing your job.'

Hawkins decided against going back to the press building. Instead he returned directly to Maryland Avenue, to write there. He began creating notes of his meeting with the presidential Press Secretary, pausing almost immediately at his recollection of Folger's reaction, staring unseeingly at the books lining his study wall. *You're going to make a lot of important people unhappy if you go into print with this.* Not a lot of important people, qualified Hawkins: *one* important person, Nelson Harriman. Who had, whatever the result of Peterson's lobbying over Nguyen van Ninh, a year of office still to run. And so still the man who had to give final permission for Ninh's entry into America.

Hawkins went to the window, gazing up towards the Capitol dome. He'd been uncomfortable with the conflict of interest over the Vietnam series. Which was nothing in comparison to this. Would Harriman be vindictive enough to block it? Bloody sure he would, if he felt sufficiently aggrieved. So what was he going to do? Hawkins remained uncertainly at the window, as the day faded into the night of Washington, darkness with a butter-yellow glow of the lights.

Unquestionably it was the best story of his career: completely international, involving the whole of Europe, not just Britain and America. Hardening a reputation already being established. There'd be wider syndication this time: probably more television appearances. Congratulatory cables. Hawkins found it very easy to recall Nguyen van Ninh, with his white-streaked hair and frail, hunched shoulders. Easy to recall, too, the weak, uneven voice: *'I've waited so long; so very long.'* If Peterson were elected, then he'd have more chance of getting Ninh out, Hawkins decided, trying to rationalise his dilemma. There was no guarantee, in fact, that Harriman would help even if he held back from writing about the collapse of the missile proposals. Just a year: not long, after already waiting for eight. *Much* easier, if Peterson were in the White House. The first responsibility of a journalist is to the newspaper that employs him, beyond all other consideration. An edict of his father, remembered Hawkins, a father who had been the first to abandon Nguyen van Ninh.

Hawkins' story was headlined 'Russian Roulette' and created the sensation he anticipated: there was another cable from Doondale, as well as from the editor. Harry Jones didn't send one but increasingly Hawkins was coming to disregard Jones as any sort of threat.

Eleanor had her head against his stomach, teasing him with her mouth to start again but he knew it would be difficult, so soon. He pulled her up and kissed her, tasting his own salt, and she said, in mock petulance, 'I was enjoying that.'

'I couldn't, not yet. You're too greedy.'

'Complaining?'

'No.'

'Did you call her?' said Eleanor.

'Who?'

'Beth.'

'No.'

'She screws everyone,' said Eleanor. 'She's got herpes.'

Hawkins pulled back, frowning at her. 'How do you know that?'

Eleanor giggled. 'I made it up, to frighten you off.'

'What do you tell John, when you're coming here?'

'I don't tell him anything: I told you, he's not interested.'

'You make it sound bad.'

'It is.'

'You going to leave him?'

She laughed, bitterly this time. 'And interfere with his presidential ambitions!'

'What about afterwards?'

She shrugged. 'I don't want to talk about it. I just want to turn off the world and be with you.'

'I've been counting,' said Hawkins. 'You're not drinking so much: not here anyway.'

'Here I've other things to do,' she said, moving down over his stomach again.

There were two telephone calls the following morning at Maryland Avenue, before he had time to leave for the office. The first was from Joe Rampallie, calling on behalf of Senator Peterson. They'd gone as high as the Secretary of State himself and got a complete rejection, so they regretted there was nothing they could do to help the exit from Vietnam of Nguyen van Ninh.

The second was from the British Embassy, asking Hawkins urgently to meet the ambassador.

Chapter Fourteen

They met in the same panelled room as before but on this occasion there were three other people with Sir Neville Wilkinson. Jeremy Abbott escorted Hawkins through the embassy but at the door of the room in which they were waiting the ambassador nodded and said 'Thank you' in a tone which clearly dismissed the press counsellor, who turned and left immediately. Hawkins was conscious of the changed attitude of Wilkinson. Before he had been avuncular and vaguely condescending: now he appeared clipped and official.

'Thank you for coming,' said Wilkinson.

'The message said it was urgent,' reminded Hawkins.

'We think it might be,' said Wilkinson. Indicating a greying, pinstriped man to his left, the ambassador continued, 'Let me introduce Herbert Smale, my First Secretary . . .' Turning to his right, Wilkinson said of a portly, bespectacled man, 'This is Harry Pearlman, Under Secretary at the State Department . . .' Of the third man, a neat, precise figure with a briefcase against the leg of his chair, the ambassador concluded, 'Edward Katzback, legal counsel from the State Department.'

The men nodded to him at each introduction and Hawkins nodded back. Wilkinson indicated a chair which put Hawkins in the centre of the group, like a witness giving evidence. He hesitated and then took it.

'Before I get to the point of this meeting I want to make something quite clear to you,' said Wilkinson. 'Nothing said in this room today must go any further. The Foreign Office in London are calling in your editor to seek the same understanding.'

Hawkins didn't respond at once, irritated at the ambassador's demeanour. Then he said, 'Until I *know* the point of the meeting, I don't know that I can give that sort of undertaking.'

The two Americans looked briefly between each other and Pearlman said, 'Mr Hawkins, we know you're a working journalist and we appreciate what that means. It's possible that the lives of several people . . . Americans . . . might depend upon your discretion. I don't think you or your editor will have any professional difficulty in meeting our requests.'

'Which makes it unnecessary imposing blind conditions before I know what it's all about, doesn't it?' said Hawkins, impressed by the determination with which he was insisting upon his independence.

This time the exchange of looks was between the ambassador and Pearlman. The American gave a nod of apparent agreement and Wilkinson came back to Hawkins and said, 'Do you know someone named Nguyen van Ninh?'

Hawkins frowned, momentarily silenced. 'He worked for my father in Vietnam,' he said. 'I established contact with him during my visit . . .' He stopped, looking at Pearlman. 'You should know that,' he said.

The American frowned. '*I* should know.'

'The State Department,' qualified Hawkins.

'I'm sorry,' said Pearlman. 'I don't understand what you're talking about.'

'He asked for my help in getting out of Vietnam: wanted me to act as sponsor. I approached Senator Peterson: I got a message today to say that it had gone as high as the Secretary of State and been refused . . .' Hawkins looked back to Wilkinson. 'I intended coming to you about it: to see if something could be done through London.'

The two Americans were in muttered consultation, Katzback fumbling with the briefcase at his side and taking out a file. He went rapidly through it, shaking his head and Pearlman said to Hawkins, 'Senator Peterson told you it had gone to State? To the Secretary himself?'

'Not Peterson directly,' admitted Hawkins. 'An immediate aide.'

'Mr Hawkins,' said Pearlman slowly. 'There's no record at all of any approach being made to the State Department involving anyone called Nguyen van Ninh.'

Hawkins shook his head, disbelievingly. 'I was told this

morning!' he said. 'About fifteen minutes before I got the call to come here.'

To Wilkinson, Pearlman said, 'Excellency, do you have a telephone my colleague can use?'

It was Smale who responded, leading the American lawyer to a side door. To the men who remained, Hawkins said, 'Why are you asking me about Ninh? What's happened?'

Instead of replying, Wilkinson looked briefly at some papers before him and said, 'What about Nicole Tiné?'

Hawkins curbed the feeling, a combination of irritation and impatience, recognising that Wilkinson was determined to dictate. He said, 'She was my father's . . . my father's common-law wife. In Vietnam. I learned of her through Ninh. He'd lost contact with her but thought she was in a re-education camp.'

'It could all fit,' said Pearlman, to the ambassador.

'What?' demanded Hawkins. 'What's this about?'

Once more there was a nod from Pearlman. The ambassador went back to the papers before him and said 'Five days ago a British freighter, the *Henry Clair,* rescued Boat People from a waterlogged vessel about forty miles south east of the Poulo Condore islands. All Vietnamese. According to messages that the *Henry Clair* sent to its head office . . .' Wilkinson looked up '. . . who fortunately acted in a responsible manner and immediately contacted the Foreign Office, two of them gave their names as Nguyen van Ninh and Nicole Tiné. Ninh named you as a sponsor for his entry permission. Tiné, too . . .'

Katzback re-entered the room, shaking his head at Pearlman. The ambassador stopped talking at the American's entry and the interruption gave Hawkins the few moments he needed to recognise what he was being told and to think beyond it. To Pearlman he said, 'You told me at the beginning that American lives could depend upon what we talked about.'

'It seems the word has got around how difficult it is for Boat People to get accepted anywhere,' said the Under Secretary. 'Ninh and the woman went beyond naming you as a sponsor. They claimed to know the location of several American POWs still being held in Vietnam . . .' Katzback offered the man the file but Pearlman shook his head against any reminder. He continued, 'According to the messages we're getting from the freighter, these two know the camp in which Francis Forest, James McCloud and Charles Bartel are being held . . .'

'What!' said Hawkins.

'Forest, McCloud and Bartel,' listed Pearlman, simply.

'They're dead,' said Hawkins. 'I've seen the film, talked to Peterson and to Patton . . . they're dead . . .'

'I've seen the film, too,' said Pearlman. 'Last night. And I've reviewed all the evidence given at the enquiry which was held after the episode. I agree everything points to their being dead.'

'Where do Ninh and the woman say they're being held?'

Pearlman gave a sad smile. 'Like I said, it seems these poor bastards know how difficult it is to get accepted anywhere in the West. We've cabled back, obviously. And they've refused to go further. They say they'll only talk to you: only trust you. And in return for telling you . . . for telling us . . . they want permission to stay.'

'Will they get it?' Hawkins demanded at once.

Pearlman shrugged. 'These two were with your father, who got back: they would have known the names of the others involved in the April mission. Known, too, the sort of publicity it had. They could be trying to con the hell out of us.'

'What if they weren't conning you?' insisted Hawkins. 'What if Forest, McCloud and Bartel are alive and they know where?'

'Are you aware of the feeling about POWs still thought to be in Vietnam?' said Pearlman.

'Yes,' said Hawkins, simply.

'They'd sure as hell get my vote,' said Pearlman. 'That's not official but if these two prove something that millions of Americans have suspected for eight years then they'd deserve to be made honorary citizens.'

'Do you appreciate now why I said what I did at the beginning of this meeting?' said Wilkinson.

'Of course,' said Hawkins.

'Do we have your assurance there'll be no publicity?'

'That isn't mine to give,' said Hawkins. 'That decision is the editor's, who's being seen in London. You've my personal assurance I'll do nothing to jeopardise people who could still be alive.'

'We want more than a personal assurance,' said Pearlman. 'Like I said, these two say they'll only deal with you.'

'What?' said Hawkins. 'I'll do anything, of course.'

'The *Henry Clair* gets into Hong Kong in four days,' said

Wilkinson. 'As well as being asked for a guarantee of silence, your editor is being asked to let you go to the island and meet them.'

'He'll give it,' said Hawkins, confidently. To the lawyer he said, 'What about Peterson's contact?'

Katzback lifted his shoulders, an uncertain gesture. 'No one has any record of any approach, like we said before. Certainly nothing has gone as high as the Secretary: I've spoken to him myself, not five minutes ago.'

'Can I talk to Peterson?' said Hawkins, remembering the embargo.

'No,' said Pearlman, at once. 'No one beyond this room; and your editor, of course.'

'You'll make all the official arrangements for Hong Kong: permission to board the freighter?' Hawkins said to the ambassador.

'The moment I hear you're going,' said the ambassador.

'I'd like to remind you again how important this is,' said Pearlman.

'You don't have to,' said Hawkins.

The message waiting on the office telex asked Hawkins to telephone London and when he did Wilsher said he wanted Hawkins to route himself through London on his way to Hong Kong. After making the reservations Hawkins called Eleanor.

'What do you mean, you can't tell me?' she demanded.

'I'm sorry,' he said.

'Why not!'

'I can't: not yet,' insisted Hawkins. Why hadn't Peterson done anything about Ninh? And then lied about it?

'How long will you be away?'

'I don't know: not long.'

'There's nothing wrong, is there?'

'No,' said Hawkins uncomfortably.

'Sure?'

'Quite sure,' he said, a lie of his own. He was finding it difficult to be sure about anything.

Chapter Fifteen

Hawkins was conscious of the growing annoyance and when he finished Wilsher said, 'What the hell did you think you were doing?'

'I'm sorry,' said Hawkins. 'I just didn't want to do anything that might endanger Ninh getting out of Vietnam.'

'I'm your editor, for Christ's sake! This newspaper employs you!'

'I made a mistake in not telling you,' conceded Hawkins. 'I've said I'm sorry. And I am.'

'What else is there that we don't know about?' said Jones. He was being very clever, letting all the criticism emanate from the editor.

'Nothing,' said Hawkins. 'You know everything now.'

'Have I got your word on this?' said Wilsher, unconvinced.

'Absolutely,' said Hawkins. He felt like a deceitful child before a headmaster and didn't enjoy the sensation.

'I think it makes sense of what we were discussing earlier,' Jones said to the editor, edging out from his muted role.

Wilsher nodded and to Hawkins said, 'Do you realise the potential of this, if it's proved that these men are alive?'

'Of course I do,' said Hawkins, irritated at the patronising note in the man's voice. He might have made a miscalculation but he didn't deserve that attitude.

'I think you need a back-up,' said Wilsher. 'It would be unfair to expect you to try to do everything alone.'

'What sort of back-up?' said Hawkins, apprehensively.

'There's going to be the need for a cover story anyway,' said the editor, not directly replying to the question. 'I think you should have some help on the spot. Harry will come out to Washington: initially we can say he's doing holiday relief.'

Shit! thought Hawkins. He didn't want anyone constantly at his shoulder; and certainly not Jones who'd use it to prove how much better he would have been as a Washington correspondent and pull rank and make every success his own and every failure someone else's: not someone else's, thought Hawkins, in immediate contradiction. His. There was nothing he could do to resist it. Resistance would have been difficult anyway because with logical objectivity he knew he did need help. Now they suspected him of with-holding on them, it was impossible to argue against. To attempt to would convince them there *was* something further he wanted to hide.

'Of course,' he said. 'It's a good idea.' The editor's office was part of the original building, with carved Victorian teak and mahogany panelling, but there was no double-glazing and the grumble of the London traffic intruded into the room.

'I'm sorry we can't back you up in Hong Kong,' said Wilsher.

I'm not, thought Hawkins. He said, 'They seem very insistent they'll only talk with me,' he said.

Wilsher nodded. 'I didn't meet any Under Secretary at the Foreign office,' said the man. 'It was the Foreign Secretary himself. This thing has got to Cabinet level. That puts a lot of responsibility upon you.'

'I understand that,' said Hawkins.

'Which reflects upon the paper,' enlarged Jones. 'If you make a mistake it won't be a personal one.'

'I don't intend making a mistake,' said Hawkins.

'What did you think of Ninh?' said Wilsher. 'No doubt about his being genuine, have you?'

Pearlman's fear at the Washington embassy, Hawkins remembered. An obvious one. 'He seemed convincing enough,' he said.

Both other men detected the uncertainty. It was Jones who responded first to it. 'What *evidence* did he produce to convince you he was Nguyen van Ninh?' he demanded.

None, remembered Hawkins: in fact everything the man had said was *different* from what he knew, from his father. So why had he believed a bowed old man and not his father? Bowed, he thought: bowed and apparently old, older than he should have been according to the documentation he'd found in the Maryland Avenue basement for a man named Nguyen van Ninh. He felt a deep hollowness in his stomach, as if he hadn't eaten for a long time.

'There wasn't any time to produce evidence,' he said, awkwardly. 'We only had one meeting and he didn't turn up for the subsequent one.'

'Didn't you try to find him again!' said Jones, allowing the incredulity into his voice.

'He said it was dangerous for him; that I'd endangered him already.'

'So you didn't bother?' pressed the assistant editor. 'You let someone come out of the boondocks, convince you he was Nguyen van Ninh and let it go at that, without bothering or trying to check further!'

'I did not involve this newspaper,' reminded Hawkins, turning the mistake into an advantage. 'I made a private request through a personal friend in Washington . . .' He hesitated at the thought of calling a personal friend someone with whose wife he was having an affair, then hurried on '. . . thus far any embarrassment will be mine, not this newspaper's. All we understand is that there are two people aboard a freighter claiming me as a sponsor and that they know the location of American POWs. If I don't believe them when I get to Hong Kong then that's the end of it. This paper is not embarrassed. Nor is anybody else. Just me, perhaps.'

It was a strong rebuttal and Hawkins saw the editor nodding, in agreement.

Refusing to be deflected Jones said, 'I hope you get better evidence than you did in Saigon.'

'I haven't embarrassed you so far,' repeated Hawkins, content with his argument. 'I don't intend to in the future.'

'I made it clear that we agreed to sit on everything only under condition of complete exclusivity, if they are genuine and there are Americans still held,' said Wilsher. 'Jesus, what a story!'

'I saw the 1975 film,' said Hawkins. 'For the book. I spoke to Peterson and to Patton, too. There didn't seem to be any doubt about their dying.'

'Which makes the possibility of these two being phoneys even more likely,' said Jones.

'We're talking about unknowns: too many uncertainties,'

said Wilsher, halting the discussion. 'We can't make sense of anything until Ray gets to Hong Kong and sees these people.'

'*Hopefully* then,' said Jones, in an over-qualification.

'At least I'll have your support and help from now on,' said Hawkins, with matching heaviness. 'It's going to be a great comfort having you watching my back.'

Wilsher looked curiously between the two men. To Hawkins he said, 'The Foreign Secretary has agreed with your liaison with the Americans but insists upon knowing everything that's going on: carbon copy, in fact. Hong Kong and the shipping company have been instructed to cooperate fully.'

Hawkins looked at his watch and said, 'I've got an onward connection in four hours.'

'The proprietor knows,' disclosed Wilsher. 'Somehow he imagines a personal connection, because it was his idea of a book in the first place. So don't worry about getting it wrong and upsetting this country or America or the newspaper. Just worry about upsetting Doondale: that's the worst thing that could happen.'

'Thanks!' said Hawkins. Christ, he needed a drink.

Nelson Harriman's grandfather had been one of the last Popularist Party politicians in the State of Texas. His father had served twenty-five years in the State legislature so any other career had been unthinkable for a boy baptised into politics practically before he was baptised into the church. It was a tradition that had gained him the ultimate in the White House and it was a tradition that had trained him to spot a political advantage with the accuracy and speed with which a hawk could spot a jack rabbit in the scrub of the Pedernales.

'From that half-assed publicity trip that Peterson was on in Vietnam!' he repeated, intent upon every fact. 'There are Americans still in captivity! Americans Peterson attested were dead!'

'That's the initial information, Mr President,' said the Secretary of State, William Keys. Recognising the importance, he had sought an immediate and private meeting with the President.

'What are we doing about it?' demanded Harriman, bent forward intently over his small desk.

'Being properly cautious at this stage,' said Keys.

'What's that mean?'

'We're not getting directly involved,' said Pearlman, the only other man in the Oval office. 'The information comes from people who were connected with Edward Hawkins: they say they'll only talk to the son. He's gone to Hong Kong.'

'What!'

The two men shifted at the President's reaction.

'There wasn't another way, Mr President,' said Pearlman defensively. 'They'd only talk to Hawkins.'

'Did we try another way?'

Keys looked at Pearlman, shifting the responsibility. 'I thought it best to keep it low key,' said the Under Secretary.

'Hawkins is close to Peterson, for God's sake! And he's a smart bastard, using it for all it's worth. Where else but from Peterson do you imagine the leak coming of the missile foul-up?'

'I'm sorry,' said Pearlman.

To the Secretary of State, Harriman said, 'Now listen, Bill. I want you to get back to the British and I want you to lean on them like hell. I want every sort of pressure out on that goddamned ship to make Hawkins realise he's among the grown-ups, not pissing about in fantasy land. If there are Americans in Vietnam, Americans I can and will get out, I've got a re-election that's going to make records. And I'm not going to let anything get in the way of it. You understand?'

'Yes, sir,' said Keys. 'I understand.'

Harriman pondered for several moments and then said, 'That's not enough: not enough by half. I'll assemble a Task Force, to handle everything. Tight, to prevent any leaks. But comprehensive. Yourselves, CIA, Defence, legal advice. Anything else can be added if and when the need arises.'

'That's an excellent idea,' said the Secretary of State, anxious to recover from his own miscalculation.

To Pearlman, Harriman said 'And I'm not trusting that goddamned Englishman. I want the pressure everywhere, like I said. But I want you to go to Hong Kong . . .' He paused, speaking as the thoughts came to him. 'I'll call CIA myself, brief the Director personally. I want someone with you who knew Vietnam . . .' There was another pause, longer this time, for reflection. Then Harriman looked up to the two men and said distantly, 'Peterson attested that they were all dead.'

Milton Snow had been awarded the Directorship of the CIA in gratitude for his work as Nelson Harriman's campaign manager and gone to Langley to confront the resentment of intelligence professionals to an amateur. Trained as a lawyer, Snow was a pragmatic realist who anticipated the attitude. His proven expertise in the campaign had been organisation, which he continued to use well, properly delegating throughout every directorate of the Agency, holding back from operational procedure until the presentation of the final analysis and then proving his intellect with subsequent queries. In three years resentment had changed to grudging admiration. The CIA is a bureaucracy in which the bureaucrats, like those everywhere, can work either for or against their appointed leaders. By the time of Harriman's request, Snow was getting cooperation, which was fortunate. It took only three hours, from the time he relayed the President's needs, to be given the name of Ben Jordan and summon the man to his office on the seventh floor of the Virginia headquarters.

Ben Jordan had been on the last CIA station in Saigon, one of the Americans plucked from the embassy roof during the helicopter evacuation. He was a quiet-voiced, aesthetic looking man who seemed more like a university professor than the accepted image of an intelligence officer, which was his advantage because he was a very good one. He had, however, disliked Vietnam, because it was clearly lost by the time of his arrival. And disliked more his enforced behaviour in it, because it was clearly pointless. Jordan enjoyed neither losing nor doing things that were pointless. He was unhappy, therefore, at the reappearance of part of his life he considered forgotten and behind him. Furthermore, he had grown to enjoy the organisational responsibility as deputy head of the clandestine division – Plans – and was disconcerted at going actively operational again, particularly upon something like this. As well as learning organisation at Langley he'd come to understand politics. Politically, to become involved in this was a bastard.

'The President?' he said, wanting to get everything clear.

'Personally,' said Snow. 'This has got to be right, from the start.'

Chapter Sixteen

The sleep was whisky-induced and Hawkins awoke foul-mouthed and thick-headed just after take-off from Tokio, squinting around in the half-darkness of the still slumbering aircraft. He made an unsteady way to the lavatory cubicle and cleaned his teeth, which was an improvement but not much. He hadn't come badly out of the meeting with Wilsher: better, perhaps, than he deserved because he should have told London although at the time it hadn't seemed so important. So he'd been lucky. But from now on luck wouldn't have anything to do with it. From now on he was trying to satisfy God knows how many people in God knows how many different ways with Harry Jones in Washington chipping and sniping and doing his best to bugger everything up. Hawkins felt frightened, more frightened than he could ever remember feeling before: even that night in London. Could there be American prisoners still in Vietnamese captivity? There'd been convincing evidence before Congressional hearings of actual sitings, long after 1975: he'd seen figures as high as 2,500 of possible imprisoned survivors. A strong lobbying league existed in Washington of families of unaccounted-for prisoners or men reported missing in action convinced there were still prisoners. But these weren't just American prisoners: these were Americans who had been on the mission with his father, who had reported them dead. And whom a careful, impartial enquiry had concluded *were* dead. Which brought him back to Nguyen van Ninh and Nicole Tiné. About the man all he had was a name and about the woman not even that. So was it a bluff – trying to con the hell out of us, as Pearlman suggested? If they'd become Boat People then they could be sufficiently desperate: they'd have to be desperate in the first place to attempt an escape down the Mekong into the uncertainty of the South China Sea. But if it were a bluff, what hope was there of their succeeding, he queried, in sudden qualification? They might gain initial acceptance with their story but that gave them no guarantee of permanent residence anywhere in the West if it were subsequently shown to be untrue. The reverse, in fact. A pointless bluff then. Another circle, with no beginning or end. Like why hadn't Peterson done anything about Ninh? And had Rampallie lied about it? That, at least, was one circle he hoped to be able to unlock.

It was still early when the plane made its unsettling landing between the over-shadowing peaks of Kai Tak airport. After the plane halted there was a special announcement for everyone to remain seated and as soon as the doors were opened immigration came aboard and Hawkins realised, surprised, that it was for his benefit. The First Secretary of the Governor's staff was with them, a thin, angular young man named George Beacher, with a fawn summer-weight suit, an Eton tie and an attitude of accusing nervousness. Hawkins followed the group self-consciously from the aircraft directly into the waiting limousine, spared any of the usual formalities.

'Kicked over a beehive with this one,' said Beacher, as the vehicle left the complex and dipped into the tunnel to take them beneath the harbour to Hong Kong island.

'How many people know about it here?'

As if reminded Beacher pressed the button inset into the arm of his seat, raising the glass partition between them and the driver. 'The Governor. Myself. Communications staff, naturally: haven't worked so hard for years. They're bound by the Official Secrets Act, of course.'

'What about the freighter?'

'Being held off the Point. Not even pilot or health authority have gone aboard yet. We're going directly to meet the Governor and then out in the pilot cutter.' Beacher smiled sideways in the car. 'Doesn't give you time for much rest, I'm afraid.'

The car emerged suddenly into the brightness of Hong Kong. Hawkins said, 'I managed to get some sleep on the plane.' He wished he hadn't panicked and drank so much. And been able to shave and bathe and change out of the concertinaed clothes. They were travelling along Connaught Road, along the harbour shore looking over to Kowloon. Tight against it but spreading out into the water like flotsam was the permanent village of junks and sampans. Floating People of Hong Kong, Boat People of Vietnam, reflected Hawkins: he supposed that was how authorities regarded them, flotsam and jetsam everyone hoped the tide would wash away and dump on somebody else's beach.

'How many Vietnamese are here in camps?' he said.

'Too many,' replied Beacher. 'Thousands. Damned difficult to know what to do with them. And it wasn't our bloody war after all.'

'Some would argue it wasn't theirs, either,' said Hawkins, quietly. It hadn't been proper sleep and he already felt tired. There was going to be a lot to do before he could think of sleep.

Sir William Makepiece was a fluttering, rubicund figure in constant, uncertain movement beneath a disarray of white hair which accentuated a face already coloured by blood pressure and further flushed by what was happening. They met in an office decorated rather than lined with books, heavy with wood and the smell of polish. There was comforting air conditioning and a view out towards the Peak. Hawkins accepted coffee with the relief of the air conditioning, wishing he was able to clean his teeth again. Bloody stupid, to have drunk like that.

'There's intense pressure about this,' said Makepiece, as if everything were Hawkins' fault. 'Intense pressure. Americans are actually sending some people. There have been instructions from Whitehall to cooperate fully. Cable from the Foreign Secretary himself.'

He wondered who or what the Americans would be: it meant, at least, that the responsibility would not be entirely his. 'The reaction is to be expected, surely?' he said.

Makepiece gave a twitching motion of his head, which Hawkins decided was an indication of annoyance. The Governor said, 'You're aware of the importance, aren't you? You're in a very delicate and awkward situation: we all are.'

'Completely aware,' said Hawkins, weary at the constant repetition. Makepiece reminded him of a tutor he'd had at Cambridge who'd tried to maintain standards but served supermarket sherry out of a cutglass decanter.

Disregarding the assurance Makepiece went on, 'You must be absolutely satisfied. *Absolutely*, you understand!'

Straining for patience Hawkins said, 'Sir William, of course I understand. I'll do nothing to embarrass anyone.' He knew the governor was still doubtful.

'It's all most unusual,' said the bird-like man. 'Most unusual.'

'If I do satisfy myself that their story is genuine what arrangements have been made?' asked Hawkins.

Makepiece looked affronted at the question. 'Arrangements?'

'Could they be brought ashore without any difficulties?' persisted Hawkins. 'Given some sort of accommodation?'

'If necessary,' said Makepiece in a voice clearly indicating that he hoped it would not be. He twitched his head towards the First Secretary. 'Beacher will accompany you, of course.'

Introduced into the conversation, the First Secretary said 'I think we should be going: the pilot cutter's being held for us.'

'Utmost care, you understand,' said Makepiece, in parting warning. 'I'm not prepared to recommend anything until I have positive proof.'

Hawkins halted, on his way to the door. 'How can you impose a condition like that?' he demanded. 'What proof do you think they could have carried in an escape boat! What tangible proof *could* there be, apart from some physical encounter.'

Makepiece's blinking became more exaggerated, at his awareness of having gone too far. 'To your satisfaction,' he said, in qualifying retreat. 'To your absolute satisfaction.'

'That was always how it was going to be,' said Hawkins, continuing on from the room.

Despite the journey weariness and apprehension, Hawkins thought the trip out to the freighter was spectacular and wished he could enjoy it more. The overhead sun hammered the flat water into gold and silver, polished by scurrying ships and sailing craft and with the island chips set in it like dull stones. Had his mother and father sailed out on days like this and thought them beautiful? he wondered. Until now he'd even forgotten that they'd once had a home here. With that awareness came another. He was wrong, quite wrong, in believing that the trunks and boxes in Maryland Avenue were a proper record of his father's life. A record, certainly: but a professional, not a private one. All the correspondence and the souvenirs and the photographs – all the memorabilia in fact – were confined to his father's working life. He'd actually found carbon copies of the man's expenses dating back to 1953 but nowhere a record of his own boarding school fees in England or letters to and from his mother, and only two photographs of them together. Nothing, absolutely nothing, about his private life. It was as if the man had planned an autobiography – which perhaps he had although he'd never spoken of it – and edited it before he even considered writing it. Which made explainable if not understandable the absence of anything mentioning Nicole Tiné or rescue efforts for Nguyen van Ninh. Explainable to him at least. He recognised anyone else would only look upon it as a personal impression.

The pilot cutter made contact with the *Henry Clair* after an hour, in a static-clogged conversation during which a rendezvous position was arranged. Hawkins stood to the side of the enclosed bridge, eyes narrowed against the heat haze, watching the black outline of the freighter gradually form upon the horizon. The almost complete flatness of the water made easy the transfer between the two vessels: Hawkins followed directly behind the pilot and the port health doctor, with Beacher coming last. They were greeted by a huge, barrel-bellied, white-bearded man in cap and reefer jacket marked with a Master's insignia who introduced himself as Lockyer and boomed 'Welcome aboard my ship. Welcome.' Professional Scotsman, decided Hawkins.

Lockyer selected Hawkins and the Hong Kong diplomat and led them to a surprisingly spacious cabin. There was a desk, meticulously neat, with a framed photograph of a blonde, smiling woman and two children, a boy and a girl, and couches attached to the bulkhead on two sides. In addition there were easy chairs, secured with rope ties to floor bolts. There were a lot of photographs around the bulkhead, predominantly of ships. Hawkins wondered if they were vessels forming the fleet or those in which Lockyer had sailed. The Captain opened a cocktail cupboard in which the glasses and bottles were secured in their individual hollows and said, 'There's practically everything.'

He shouldn't, Hawkins accepted, but he needed it. It was a large measure with no water offered and he was grateful, feeling the alcohol move through him with familiar, welcoming fingers.

'First Officer is dealing with the pilot and health clearance,' said the Master. 'You wouldn't believe the cable traffic this thing's caused.'

'I think I would,' said Hawkins.

'I've been ordered to cooperate completely,' said the Captain. 'Government instructions, as well as the office . . .' Lockyer looked seriously between the two men. 'I want to make it clear that I understand you've got your responsibilities, like I've got mine,' he said. 'But aboard this vessel I'm in command: I must be satisfied, about everything.'

'We understand your position fully,' said Beacher, smoothly. He'd politely accepted gin but put it untouched on the table beside him.

'How do you want to proceed?' asked Lockyer.

Hawkins was aware of Beacher looking to him, for a decision. 'From you first,' he said. 'I want to hear about it from the very beginning.'

Lockyer put himself in a chair facing them and Hawkins realised what a big man he was, so fat that for comfort he had to sit with his legs splayed. 'Not a lot to it,' said the Master. 'The Poulo Condore islands are just off the tip of Vietnam so we always give them a wide berth. I was about forty-five miles south by south west on the fifteenth when we picked up something on the radar. No superstructure to give us any identification so we thought at first it was a hazard. Maybe a wreck or just a collection of floating driftwood and debris. Checked it, of course. When we got near enough we saw it was a dismasted hulk, just drifting. It was virtually waterlogged: shipping water at every movement. We put a boat over and found there were eight people aboard. They were in a pretty bad way. Dehydrated through lack of water, a child of maybe twelve almost dead, from lack of food. We got them aboard: nothing we could do really, except water and feed them. After about three days the second officer said one of them wanted to talk to me. Told me his name was Ninh, that he'd once worked for a very important Englishman. Mentioned Edward Hawkins . . .' The Captain paused, nodding towards

Hawkins. 'And then you. Said he'd been with you in Vietnam very recently, which made me think he was lying . . .'

'I was there,' said Hawkins. 'He wasn't lying.' I hope, he thought.

Lockyer sipped his drink and continued, 'Said you would act as sponsor for his acceptance into the West. And for a woman picked up with him, Nicole Tiné. And then he announced he knew where there were some Americans, still held in captivity.'

'Did you ask where?'

Lockyer gave Hawkins a hurt look. 'Of course I did. He wouldn't tell me. He said he'd talk only to you.'

'Did you speak to the woman?'

Lockyer nodded. 'Ninh said she'd just been released from a camp and it certainly looks like it. She's very frightened: she won't talk without Ninh being present and always defers to him. Don't find it difficult at all to believe that she's had a hard time. Fit enough, though.'

'What's happened to them?' asked Hawkins.

Lockyer frowned. 'I don't understand the question.'

'Have they remained with the others you picked up? Or been separated?'

'As soon as I began to get head office reaction I separated them, which wasn't easy in a ship like this.'

'Do you know if they've talked to the others you picked up about the Americans?'

'No,' said Lockyer.

'They were specific about the names, Forest, McCloud and Bartel?'

'Ninh was,' said Lockyer. 'I don't remember the woman identifying them: she just talked about Americans.'

Ninh would have known the names from his father, thought Hawkins. All the repeated warnings and the paraded doubts and the expressed reservations crowded into his mind. 'Tell me about the ship. In your professional opinion as a sailor were they genuinely shipwrecked? They couldn't have been towed there, for instance: left for you to find them?'

Lockyer shook his head, as if he had difficulty digesting the question. Then he said, 'The child I told you about: it died. So did an old woman: I think they were related although I can't properly establish that. They were *all* suffering from exposure. Had been, for days. There's no way it could have been arranged, like you suggest. It wouldn't have worked, because of the currents. What would be the purpose, anyway?'

'I don't know,' admitted Hawkins. 'I'm just trying to anticipate things that might be asked later.'

'What proof is there of identity?' intruded Beacher. 'Passports, documents, things like that?'

'None,' said Lockyer at once. 'They've just got the clothes they're wearing, nothing more. Certainly no documentation.'

'Do Boat People often have documentation?' Hawkins asked the diplomat.

'No,' admitted Beacher. 'I just wondered if these two had.'

'I've no proof,' reinforced Lockyer. 'The names I gave to London were the names they gave me, verbally.'

Proof, thought Hawkins angrily: there was never any damned proof.

'Another drink?' offered the Captain.

'No thank you,' refused Hawkins, reluctantly.

'Anything else I can help you with then?' said the man.

He wished there were, thought Hawkins. He wished, somehow, something would happen to make it easy and take away the pressure: make it so he could decide, one way or the other, and get suspicious, distrustful people off his back and get Jones off his back: everyone off his back. He didn't want to be here, exposed like this, he thought, in a surge of self-pity. He said, 'I suppose we'd better see them.'

Lockyer picked up an internal telephone and said 'Bring them in,' without appearing to dial. Within minutes there was a knock at the cabin door which was opened to Lockyer's shouted permission by an officer shepherding two people into the cabin ahead of him. Ninh entered first, with stumbling, frightened hesitancy. He appeared to be wearing the same clothes as that day upon the Saigon dockside but much dirtier, the trousers shining with grease and wear and bagged around his skinny legs, shirt more frayed than Hawkins remembered and smeared with dirt. When Ninh saw Hawkins the Vietnamese tried to smile but it was a difficult expression because his lips were still blistered and split from exposure in the water. The ribbon of strangely white hair hung forward over his face, like a flag of surrender.

Ninh's progress into the room showed the woman behind him and Hawkins gazed at her in bewilderment. He had expected her to be Asian-featured, another Vietnamese in fact. But she was completely European, the only indication of her mixed parentage the dark sallowness of her skin and the deep, black-brown eyes. Her apparent age surprised him, too, although it shouldn't have done because Ninh had told him during their Saigon encounter. Although he'd known, Hawkins had created a mental imagery of someone as old as his father but she clearly wasn't. And the youngness hadn't been taken away from her by whatever ordeal – or series of ordeals – she had undergone, as they had marked the wizened, bowed Ninh. There were sores around her mouth from the open boat trip and her skin was still patchily puffed and red from the sunburn but Hawkins recognised it would be easy for Nicole Tiné to be beautiful again. They stood in the middle of the cabin, eyes moving around them from beneath slightly lowered heads. Almost in unison, both gave a deferential Asian bow, lowering their heads further.

No one – neither the Europeans nor the refugees – appeared sure of what to do. Realising that the Captain and Beacher were deferring to him Hawkins said 'Hello,' more to Ninh than to the woman, wondering incongruously where he'd read that people never found words sufficiently important for important occasions.

'We have caused you trouble?' said Ninh, in his thin, reedy voice.

'No,' said Hawkins, thinking that could be the understatement of this or any other year of his life. 'You haven't caused me any trouble.' He thrust out at last, wanting to make some physical gesture. Ninh hesitated and then took the offered hand and when he released it Hawkins stayed uncertain whether to make the same gesture to the woman. Stuck with the inadequate word he said 'Hello' to her and reached out again. Her hand was soft and feminine.

'Thank you,' she said. 'Thank you for coming.'

'I knew you would,' said Ninh immediately. 'I knew we could trust you.'

Would it be possible for these crushed, emptied people to attempt any sort of deception? Hawkins asked himself. Taking the lead Lockyer and Beacher appeared to expect him to, Hawkins said in sudden awareness: 'Please! Please sit down!'

He tried to pull one of the chairs closer to Nicole, forgetting it was storm-tethered to the floor. It jerked out of his hand and the woman sniggered at his clumsiness and Hawkins smiled too, glad at the slight relaxation between them. She sat where he indicated, and Ninh lowered himself just as obediently into the adjoining chair.

'There is a great deal I have to know,' said Hawkins. Indicating Beacher, Hawkins said, 'This man is from the government: the British government.'

Ninh nodded, a man proved right and said, 'I knew we should wait. I knew you would arrange our entry.'

Hawkins shook his head towards them, intending it as a warning. 'I have to know a great deal first,' he repeated. 'I must know what happened: everything that happened.'

Ninh looked down into his lap, preparing himself. 'They found out,' he said. 'The authorities found out about our meeting. I knew they would but I had to come, of course. I wouldn't have known what you were going to do otherwise . . .' He smiled up. 'I always knew something would be done,' he said, offering the trusted litany like a special souvenir in a child's collection. 'He promised. Your father promised.'

'When did he promise?' said Hawkins. It broke the man's narrative but there was so much to find out.

Ninh frowned, as if it were an overly-simple question. 'Always,' he said. 'Always if the war was lost that he would help me to get out.'

'And you,' Hawkins said to the woman. 'Did he promise to get you out?'

'Of course,' she said, treating the question as Ninh had. Her voice had the blur of a French accent.

'Why didn't he?' demanded Hawkins. 'Everyone knew the war was ending. The talks had gone on in Paris for months: the American withdrawal was public knowledge. When it became obvious that it was over why didn't he get you both out?'

'I had a mother, in Dalat,' said Nicole.

'I know about your mother: about Dalat,' said Hawkins. 'Your father was French?'

'Yes,' she said.

'Under French law – the law they extended to their colonies and which was still in force – that gave your mother and yourself the right of French citizenship.'

'My father was dead,' said the woman. 'My mother didn't

know France, didn't know anywhere but Vietnam. She said it wouldn't be as bad as everyone feared: that we should stay.'

'You believed her,' said Hawkins.

'I couldn't leave her. It is not our culture for children to abandon their parents,' she said.

'What about you?' said Hawkins, coming back to the Vietnamese. 'Why didn't you get out in time?'

'The papers did not come,' said Ninh. 'He tried to get them. Paid money, which was the way, but they did not come. He said I was not to worry: that he would get them, in the West.'

'From whom in the West?' intruded Beacher. 'English? American?'

Ninh shrugged. 'It was never said.'

No proof! thought Hawkins: nowhere was there proof. Worse than an absence of proof, it was so vague. To Nicole he said, 'How long were you with my father?'

She smiled, in recollection. 'Two years.'

'There was a baby?'

The smile went, as quickly as it had come; instead her face closed in an expression of sorrow. She nodded, not speaking for several moments. 'A girl,' she said. 'Elian: it was the name my father called my mother.'

Christ, why hadn't he taken that second drink, thought Hawkins, forcing himself on. 'Why didn't you marry?'

Nicole stared unflinchingly at him. She said, 'He did not ask me.'

The sea was so flat there was no creak of motion in the cabin. There was a sound of movement as the heavy Lockyer shifted in his chair, behind the desk, but otherwise the room was quiet and still.

'How did you meet?' asked Hawkins. His voice was uneven.

'The Cercle Sportif: it was a club.'

'How old was he?'

The woman made an uncertain gesture. 'I do not know: not the actual age. Fifties, certainly. It was a joke between us. He accused me of seeking a father figure.' There was another smile, of apparent tender recollection.

Dear God, thought Hawkins desperately, when was there going to be something he could recognise, be sure about? His father had to be fifty-two, maybe fifty-three when they met. Near enough, he supposed, but still vague. Like everything else. Abruptly he demanded, 'What hand did he write with?'

The woman frowned and said, 'His left. Strangely.' She cupped her arm in an encircling way, holding an imaginary pen which would have written above, not below the words. 'Like that,' she said. 'I have seen Americans write like it, but never a European.'

'Yes,' admitted Hawkins. 'He did.' She could have learned it from Ninh, he supposed, if Ninh were who he was supposed to be.

'Why don't you believe me?' Nicole asked, quietly.

'I do believe you: want to believe you,' said Hawkins anxiously. 'A lot of other people have to believe you.' Circles within circles, he thought: enclosing him.

'You *know* who I am,' said Ninh, pleadingly.

Despising himself Hawkins said, 'I know who you told me you are.'

'You said your father sent you.'

'Did you?' demanded Beacher, from beside him.

Unwilling to admit his deceit and despising himself for the cowardice Hawkins said, 'He left some papers: that's why I came to your house.' How did he know it was the house of a man called Nguyen van Ninh?

'What papers?' said the Vietnamese.

The tightest circle of all, thought Hawkins: now he was being asked to justify himself! Truculently he said, 'A question for a question: what did the correspondents call the daily press briefings, at the MACV building?'

Ninh's expression was of patronising resignation. 'The Four o'clock Follies,' he said.

'How were they given?' He wasn't going to be patronised, Hawkins decided. He wasn't going to be patronised or cheated or made to look a fool. He was going to get it right, whatever it was.

'Given?' said the Vietnamese.

'Given,' insisted Hawkins, refusing to go further.

Ninh looked at the woman, then around the cabin. 'A big room,' he remembered. 'Maybe enough for a hundred, two hundred people if necessary, which it rarely was because the briefings were a joke. Nobody believed the body counts or the claims made.'

All of which his father had told him, recalled Hawkins;

hopefully. But all of which had been published and which was known, as well. 'Given?' he insisted, doggedly.

Ninh stared at him nonplussed, then in sudden recollection said, 'Two languages: always Vietnamese first. Then American.'

He was getting close but he still had to be careful, Hawkins realised. 'So?' he said.

The giant Captain moved again, more noisily this time, a large man unusually confined in a constricted space. Beacher sat immobile, only his eyes travelling between question and answer. The woman's hand was exploring the damage to her face, probing the splits and protuberances.

'I was interpreter,' said Ninh, in belated realisation. 'We would both attend, when it was necessary. I would listen to the Vietnamese and he would listen to the English, like I would. If there was a different answer I could tell him, so he could challenge it.'

To the Captain Hawkins said, 'Do you have paper? A pen?'

Grateful for the opportunity to move Lockyer rose, bringing both towards them, changing direction when Hawkins indicated Ninh with a nod of his head.

'What else was joint?' said Hawkins.

This time the awareness was quicker. 'Signature,' said the man. 'On the accreditation.'

'Write for me,' invited Hawkins. 'Write both for me.'

Ninh hesitated, looking for a rest and Lockyer said 'Use my desk.'

The Vietnamese rose, doubtfully, and went towards it, putting the paper down from where Lockyer had taken it. He remained upright, bending and from behind Hawkins said, 'Don't stand: please sit. Sit as you would have done, when you signed.'

Ninh looked from the Englishman to Lockyer. The Captain nodded and gestured to his own chair: it engulfed Ninh, with its size accentuating the man's frailty. Pen in hand the Vietnamese hesitated, appearing to concentrate. Determinedly – sharply – he bent over the paper: one inscription took longer than the other. He looked up, enquiringly. Lockyer responded, retrieving the paper and carrying it across the short distance to where Hawkins remained sitting.

From an inside pocket of his travel-crushed jacket Hawkins

took two documents, the accreditation application and the pass itself. Against them he carefully placed the paper upon which Ninh had just written. With the Vietnamese he was insufficiently familiar to make a proper comparison: it seemed identical to his unpractised eye. The English version was unsteady, as if it had been written with laborious difficulty. It still was a remarkably good match to the faded inscription, beneath the preserving yellow plastic, upon the documents he'd recovered from the Maryland Avenue trunk devoted to Vietnam. Wouldn't someone unused to a written language like English – further unused for eight years – have responded with the painstaking care that Ninh had shown? wondered Hawkins. He offered the comparison sideways, to Beacher.

'There's something further,' said Ninh, indicating the documentation. 'There's provision for a photograph: my picture should be there! What more proof do you need than a picture?'

'There isn't a photograph,' conceded Hawkins, who'd seen the provision upon the pass and been unable to understand why one wasn't there.

'There should be,' insisted Ninh, with desperate defiance.

'I recognise that.'

'What is it you want?' said Nicole, quiet in comparison to the man alongside her.

'Proof!' exclaimed Hawkins. 'Give me proof!'

'Of what?' said Ninh.

'Anything,' said Hawkins, his desperation close to the other man's. 'Give me proof. Proof is what I want.'

The woman looked towards Ninh, who nodded. Nicole was wearing a wear-ridged, soiled smock that ended just over her hips, covering an equally crumpled but even more soiled canvas skirt, blotched and puddled with stains. Without embarrassment – which Hawkins later found surprising – the woman lifted the top sufficiently high to show she was not wearing a bra. Freed of the encumbrance she picked at the waistband of the dirty skirt, worrying at the threads with her thumb and forefinger. It took a long time because her nails were broken and short and yet again there was complete silence in the cabin: not even the uncomfortable Captain Lockyer changed position this time. The stitching lengthened under the attack and Hawkins saw the doubled-over waistband was gradually parting, disclosing a narrow, channelled envelope. She tried twice to get inside and failed, picking further at the binding. She succeeded on the third attempt. She felt inside, extracted something in the concealment of her cupped palm and offered it, not to Hawkins, but to Ninh. There was another nod of permission from the Vietnamese and she moved her hand sideways, towards the Englishman.

Hawkins came forward from his rope-secured chair to get the offering, staring down at what he held in his hand. They were stained, a smear of colours, yellow but predominantly black but still clearly identifiable, which after all was the function of the dog-tag. In the order she had given them to him McCloud's was first, then that of Colonel Forest and finally Bartel's.

'They asked for help,' she said pointedly. 'Like we're doing.'

Drained and aching Hawkins held them sideways towards Beacher. He said, 'They're American military ID: Official identification.'

'I know what they are!' said Beacher, impatiently. To the woman he said, 'Where did you get these?'

'They gave them to me,' she said. 'The American, Forest.'

'Where?'

'The last camp I was in.'

'Where?' repeated Beacher.

Before she could answer Ninh said, 'You will let us in? Give us permission to stay?'

'It's enough,' said the relieved Hawkins, to the other Englishman. When Beacher remained uncertain, Hawkins said 'You can't ignore the dog-tags, for Christ's sake!'

'I'm not ignoring them,' said Beacher. To Lockyer he said, 'I think they should be released into my jurisdiction. Will that satisfy you?'

The huge man came nearer, looking down at the identification discs which Beacher still held. 'Poor bastards,' he said, distantly. 'Eight years!'

'Cognac?' offered the British ambassador.

'Thank you,' accepted Harry Jones. 'And thank you for the lunch. It was excellent.'

'Why don't we take it in the study?' said Sir Neville Wilkinson.

Jones followed the diplomat from the dining room, accepted

the brandy bowl and sat down. 'I wasn't aware you knew Lord Doondale,' he said.

'Harrow together,' said Wilkinson. 'Got a damned fine shoot near Auchterarder. Know it?'

'No,' said Jones, delighted with his discovery.

'I suppose we should hear from Hong Kong sometime today,' said the ambassador.

'Providing there are no difficulties,' said Jones, dribbling out the beginning of the doubt he wanted to sew.

'It there's any difficulties with them then that'll be the end of it,' said Wilkinson, misunderstanding. 'Probably save us all a lot of trouble.'

'I didn't mean with the Vietnamese,' said Jones. 'I only hope Hawkins can manage it.'

The diplomat looked surprised. 'Seems a very capable chap to me. Made quite an impression here in Washington.'

Careful, thought Jones. Exaggerating, he said, 'Lord Doondale still felt that someone senior should be involved.'

'Quite!' said Wilkinson. 'Glad to have you here. The whole thing's damned tricky.'

'Which is why I think we should keep in close contact,' said Jones.

'Absolutely essential,' agreed the ambassador and Jones smiled, content with the success of the meeting.

Chapter Seventeen

Having so recently wished to be spared the responsibility Hawkins realised he should have been relieved to find the Americans waiting when he got back to Hong Kong. Instead his attitude was of resentment, at the thought of not being trusted to find the truth, which he recognised as illogical and maybe not a consideration in their minds anyway but still an attitude he couldn't avoid. Frightened though he'd been – and still was – he'd liked being centre-stage and didn't want to be relegated to a subsidiary role, which was as illogical as feeling resentful. More circles.

From the size of the table and the arrangement of the chairs around it, Hawkins supposed it was a conference room. Ninh and the woman sat at one end, tight together for reassurance and Hawkins put himself next to them, wanting to provide the same feeling. Were the signature and the dog-tags enough? Impartially he supposed he should seek more but he was reacting to instinct not impartiality and instinct told him they were genuine. They'd held back on the returning cutter, unsure of him now and Hawkins regretted it. Pearlman and Jordan – who had been introduced as 'someone in the government' – were opposite, the comparison signatures and the dog-tags on the table in front of them. Beacher remained standing and Makepiece hovered, uncertain whether to sit or stand.

Jordan picked up the tags, smiled across at Hawkins and said, 'They look OK to me.'

'What about the signatures?'

'Leave that to the experts,' said the American. He smiled gently at the refugees and said, 'Thank you: thank you very much for what you've done.'

'We are allowed to stay?' said Ninh, persistently.

'A little more,' avoided Pearlman. 'We'd just like a little more help. Where is the camp? Where did you get these tags?'

The couple looked at each other and Hawkins guessed they were considering bargaining, the promise of entry in return for the name. '*Then* we'll be allowed to stay?' said Ninh, confirming the thought.

'I'm sure there'll be no problem,' said Pearlman.

'That's not an answer,' intruded Hawkins.

The Under Secretary frowned at him and said, 'We appreciate what you've done, Mr Hawkins. We'll take over now, OK?'

Hawkins flushed at the dismissal, aware as he did so that the Vietnamese was speaking to him.

'Is it all right?' said Ninh. 'Is it all right for us to answer?'

Given his response Hawkins looked back to Pearlman and said, 'Is it?'

Pearlman's mouth was a tight line, his face white with anger. 'I said there'll be no problem.'

'No,' corrected Hawkins. 'You said you didn't *think* there'd be a problem.'

Pearlman looked to the Governor for assistance and Makepiece said, 'I really don't know why this attitude has arisen: don't know at all. The instructions are to cooperate.'

'*Your* instructions,' corrected Hawkins again. He realised it was getting out of control, which was stupid: that wouldn't help Ninh or Nicole. And it wouldn't help him, either. He said, 'They're stateless: nowhere to go. Wouldn't you seek a guarantee, in their circumstances?'

'The request, from the ship, was for you to act as sponsor?' said the quiet-voiced Jordan.

'Yes,' said Hawkins.

'Are you prepared to be, upon what you know?' demanded the man. 'Are you prepared to accept their bone fides upon what you know already?'

More responsibility, Hawkins recognised at once. It wasn't the Americans being asked to give an undertaking now. He was conscious of Ninh and the woman gazing at him but couldn't respond to their look.

'Are you?' pressed Pearlman.

Trapped, Hawkins said shortly, 'Yes. Yes I am.'

To the couple Pearlman said, 'You will be allowed into America under the guardianship and protection of Mr Hawkins, who's agreed to undertake full responsibility for you.'

Which was like being awarded temporary custody of unwanted children, Hawkins recognised. It gave him a burden and provided them with no protection at all, only for as long as he remained in Washington. Would Britain allow them in if he were moved back to London, which he expected to be some time? There was a professional advantage, Hawkins recognised, on balance: he couldn't be dismissed, as Pearlman had tried to dismiss him earlier.

Ninh and the woman were both smiling, uncertainly: Hawkins saw Nicole feel out for Ninh's hand. 'Thank you,' said Nicole.

'The camp?' Jordan reminded her. 'Where was the camp in which the Americans gave you these identity discs?'

Nicole looked enquiringly at Hawkins, who nodded. 'Can Tho,' she said. 'Very near Can Tho.'

'When?' said Jordan, forward in his seat.

She made an uncertain movement with her shoulders and said, 'Six months ago. I was moved from Can Tho to Pleiku, to be released.'

To Ninh, Jordan said, 'You? Did you meet them?'

The Vietnamese shook his head. 'I was at Can Tho but earlier. There were no Americans held when I was there.'

Coming back to the woman, Jordan said, '*How* did you get them: the discs?'

'Americans are kept in a separate compound,' she said. 'I had to act as an interpreter because I speak English and Vietnamese. When they learned I was being sent to a release camp, they gave them to me. They said to help . . . they wanted help.'

'Just three?' pressed Jordan.

She shook her head. 'Five when I first went to Can Tho. One died.'

'Do you know his name?' asked Jordan quickly.

'I think it was Lindsay,' said the woman. 'There is a grave, with a cross.'

From a bulging briefcase at his side, of which Hawkins had been unaware, Jordan took a file. He brought out a booklet and paged through it rapidly. He looked up to the woman and said, 'Howard Lindsay?'

'I did not know the first name,' she said.

'*Why* did they give you the tags?' he said.

'I told you: they wanted help. They said that the discs would help me, too. That I was to approach someone from the West and get help for us all.'

'We've got three tags,' said Jordan. 'A man called Lindsay died. What was the name of the fifth man?'

'Page,' said Nicole. 'He was a black man.'

Jordan went back to his booklet and Hawkins thought admiringly that the American was a good questioner. Jordan came back to the woman and said, 'Wilbur Page, missing after a search and destroy mission near Hue in January, 1974?'

She shook her head. 'I never spoke to him: he couldn't speak . . .' She put her hands to her head. 'He was unwell. There is interrogation and questioning with drugs: they are all unwell, but Page is worst. He wouldn't mix with the other Americans; just lay in his cot or sat in the corner of his hut. He always wanted darkness.'

Jordan looked briefly to Pearlman, then back to Nicole. 'I have some photographs,' he said. 'Can you look at them and see if you recognise any of the men?'

As he spoke Jordan took twelve large prints from another file and dealt them out towards her, reverse from himself so that she could see them from where she sat. Nicole stood, however, gazing along the three lines that Jordan created. Her head moved, positively, from picture to picture, from the top left to the bottom right, without any sign of recognition. Ninh went to stand, to join her, but Jordan raised his hand in a stopping motion. It was utterly quiet, like it had been quiet in the cabin, and Hawkins felt a surge of anxiety. Come on! he thought. Come on!

'These pictures were taken of men when they were well,' said the woman uncertainly.

'Yes,' agreed Jordan.

Nicole shook her head and Hawkins' anxiety increased. Don't let her fail now! he thought: this is *the* test.

'I don't think they are here,' she said. 'I can't recognise any of the men in these photographs as being those I saw at Can Tho.'

Hawkins was against his chair edge, willing a different reaction from her, and felt a sink of emptiness.

Jordan smiled, in some private triumph, collecting the prints like the gambler who'd won the pot. He tapped the pictures into order and replaced them in the folder. Nicole looked worriedly at Hawkins and turned, to return to her chair. Jordan said, 'How about these?' and began dealing out a fresh selection.

Nicole went back to the table, looking at the new photographs with the earlier intensity. Please, thought Hawkins: please! She examined the images quite impassively: Hawkins knew she realised the importance and was surprised at her equanimity. She stopped at the third from the left on the top line and said, 'Again this is of a well man but I think it is McCloud . . .' She indicated the last in the line and said, 'That is Forest: I'm sure of that.' She put her hand to the left side of her face and said, 'He has a scar now. Here.' Her head moved on and at the bottom line she said, 'I think that is Bartel . . . he has no hair, not any longer . . .'

Jordan sat back in his chair and said, 'Thank you, Madame Tiné. Thank you very much indeed.' To Pearlman he said, 'Right, every time.'

The State Department official smiled and said, 'I think we should go back to Washington. All of us.'

The specially created Task Force was still sufficiently small to be accommodated in the Oval Office. In addition to the CIA Director and the Secretary of State there was Richard Godsell, Secretary for Defence, General Cornell Bell, Chief of Staff of the Army and William Erickson, the Presidential counsel. Harriman sat at his small desk, repeatedly jabbing at the same spot in the blotter with an ornate paper knife, listening without interruption to Milton Snow's account of the Hong Kong meeting.

When the CIA chief finished, the President said, rhetorically, 'What the hell have we got here!'

No one made the mistake of trying to respond. Harriman looked up from the destroyed blotter and announced, 'I want more. Independently I want more. What satellites have we got?'

'We're well placed,' responded Snow, who had anticipated

the question. 'We've been monitoring the Vietnamese incursion into Kampuchea for months.'

'Good,' nodded Harriman. 'If it's necessary, shift it. Bring it directly into orbit over Can Tho. Let's see if there's aerial evidence of this camp.'

'We'll have to regularise in some way the entry of the two into America,' warned Erickson. 'From what Pearlman said from Hong Kong the Englishman has taken responsibility but that still doesn't make them permissible, according to the law.'

'Couldn't immigration issue some temporary permit?' said the President.

'Easily,' said Erickson. 'I thought the concern was to minimise the number of people involved, to cut the risk of leaks. We get involved with clerks and typists and God knows what bringing immigration in.'

'You're right,' said Harriman, reflectively. He said, 'I'm not going to be stuck with a couple of dummies if they're feeding us bullshit.'

'It could be by a Presidential order,' advised Erickson. 'It's not normal but it's possible. That way it restricts knowledge to us in this room and it makes it your personal decision, something you can rescind with a signature any time you want.'

Harriman nodded, considering the suggestion. 'And later it would also publicly indicate my personal commitment if everything checks out, wouldn't it?' he said, in another of his near-private conversations.

'Absolutely, sir,' said Erickson.

'That's how we'll do it,' decided Harriman. 'It's a good two-way bet.' Then he said, 'Doesn't the woman have right to French nationality, if she had a French father?'

'That was the French system, as I understand it,' said Erickson.

To the CIA Director, the President said, 'See if you can check it out through Paris. Let's see if anyone knows about her, in fact.'

'What about a feasibility study for rescue?' said Bell, wanting to be ahead of the demand.

'Everything,' said Harriman, agreeing with the Army chief. 'I want every option: air, sea, the lot. If we can get confirmation I want to be able to move like that . . .' He snapped his fingers.

Chapter Eighteen

When it became obvious Washington intended it to be an exclusively American enquiry – which should hardly have been a surprise anyway – Sir Neville Wilkinson intervened, insisting upon British representation and in London the American ambassador was called to Whitehall, for that insistence to be personally repeated by the Foreign Secretary. That gained the admission of Herbert Smale, the First Secretary at the British embassy who had been involved in the original discussion. From London, too, Lord Doondale and David Wilsher separately reminded the Americans that their undertaking not to publish had been given in return for complete and continuing cooperation and further that at American invitation in Hong Kong their representative had made himself personally liable for the refugees. The dispute was finally settled by the surprisingly forceful refusal of Nguyen van Ninh to talk to anyone about anything without Hawkins being present. When the Americans allowed the concession to Hawkins, Harry Jones insisted upon being included.

The hearing was convened at the CIA headquarters at Langley, to maintain the necessary and absolute secrecy and because Jordan, a CIA Division Director, initiated the investigation in Hong Kong.

When he entered the third floor chamber, overlooking the wooded valley of the unseen Potomac, Hawkins realised it was going to be a legally-orchestrated hearing. Edward Katzback, the State Department lawyer who had been involved in the original meeting at the British Embassy on Massachusetts Avenue was there, accompanied by the Presidential counsel, William Erickson. Peter Paulson, the Agency counsel, completed the legal panel. Pearlman and Jordan attended and there were three unintroduced men whom Hawkins naturally assumed were CIA officials.

There was a lot of security before they gained admission, increasing the apprehension of Ninh and the woman and as he shepherded them into their directed seats Hawkins decided that Ninh's refusal to talk only in his presence was motivated not by any hidden strength but by abject nervousness.

'We want to thank you for coming,' Katzback said to the couple, adopting the role of chairman. 'There is much more we'd like to know but we want you to appreciate that this is in no way any sort of trial or enquiry: it's just a group of people trying to get answers to questions. OK?'

Ninh and Nicole made vague gestures of understanding and Hawkins was relieved at the apparent kindness from the balding, fatherly man. Ninh was questioned first and began badly, over-eager to please, responding to questions before they were completed and causing misunderstandings. After several delays, for correction, Katzback warned him about being too quick with his replies and although the rebuke was made with the earlier kindness there was at once a pendulum swing, the Vietnamese humbled by correction. It took a long time for an understandable chronology to begin to emerge. But at last it did.

Ninh recounted his employment by Hawkins' father and his arrest and appearance before the investigatory committee in Saigon, which he dutifully called Ho Chi Minh City. There had, he said, been a series of re-education camps. The first was at Ban Me Thuot, an agricultural settlement. After nine months he had been transferred to the north, to Ninh Binh. That had been more of a political encampment, with lectures that lasted most of the day with work having to be done at night. It was in Ninh Binh that he first heard suggestions of American prisoners still being held after the war's end but he never saw any there. After eighteen months he was moved to his third camp, Pleiku, and then to Can Tho. It had just been established when he was sent there and the work had been the hardest because they had to continue the construction, manually hauling timber and building materials from the river in which they had first had to work, waist deep, creating a jetty construction for the later arrival of the delivery boats. It was here that the Pleiku-contracted malaria had worsened. There had been no medication or treatment and apart from the times when the fever had been at its height, he had not been excused from work. It was planned as a large camp and different from others he had been in, each of which had been modified from existing barracks or structures and not purpose-built. Despite the labouring, the political lectures had been maintained. He'd again heard stories from other prisoners of surviving Americans still in the country but once more he had never encountered any himself. The camp was still incomplete when he left, for a government-imposed residential period on another agricultural commune hamlet near Hanoi. He had only been allowed to return to Ho Chi Minh City eighteen months previously; and it was there that the son of Edward Hawkins had found him. He had feared the contact and within a month learned that he was once again under investigation. He was already making plans to flee when Madame Tiné made contact with him.

Throughout Ninh's evidence Katzback led the prompting, occasionally aided by the CIA counsel. It was only when Katzback looked invitingly either side that the Presidential counsel, Erickson, responded, snapping eagerly forward.

'You knew of the mission to Chau Phu, to rescue the orphans?' he said, and Hawkins was at once aware of the change of tone.

'Of course,' said Ninh. 'It became a famous event.'

'And the names of the Americans involved?'

'I had forgotten them until I met Madame Tiné again.'

Erickson's head came up sharply from his yellow legal notepad. 'Forgotten them until you met Madame Tiné again?' he echoed.

'Yes,' said Ninh.

The Vietnamese had relaxed as much as it was possible for him to relax: now the twitching apprehension was returning.

'I find that interesting,' said Erickson. Back to his notes he said, 'Did you fear punishment from the authorities, at the end of the war?'

'I knew it was a possibility.'

'What did you do about it?'

'Do?'

'Many Vietnamese attempted to leave the country.'

'Mr Hawkins promised to help me get out.'

Erickson looked sideways, to where Hawkins was sitting, then back to the Vietnamese. 'Did he try to help?' asked the lawyer.

'I was arrested very quickly: held in prisons for six months, before I went before the committee,' said Ninh. 'He promised. I know he would have kept his promise. His son came.'

'How did you feel about that?'

'Grateful,' said Ninh at once, smiling to Hawkins. 'I knew I would be helped.'

'But you said you feared contact with Mr Hawkins would lead to renewed difficulties with the authorities,' seized Erickson. 'What was there to be grateful about?'

'I thought it would be arranged at the first meeting: that there wouldn't be any wait. It was only afterwards that I became frightened.'

'How long were you in prison?'

'Six months,' said Ninh.

Erickson smiled up at the man. 'Thank you for the accuracy of your reply,' he said. 'I meant how long were you held, in total, in prisons and camps and resettlement places, before being set completely free.'

'I was arrested in May, 1975. There were six months in prisons, in Ho Chi Minh City and Vung Tau. Then, after I appeared before the investigatory committee, came the camps I've already spoken about. I left Can Tho in May, 1980, for resettlement near Hanoi. I was there two years.'

'Almost seven years in prisons, camps and resettlement areas?' said the Presidential counsel.

'Yes.'

'A very long time.'

'Yes.'

'A hard time, with much work and in bad health?'

'Yes,' recited Ninh.

This was becoming the sort of demanding interrogation the other lawyer had promised it would not be, Hawkins thought.

'How did you live when you got back to Saigon: what job did you have?'

Ninh hesitated, looking to Hawkins then back to the lawyers. 'The government,' he said finally.

'You worked for the communist government of Vietnam!' said Erickson, his voice pitched to show innocent surprise.
'Yes,' said Ninh.
'Doing what?'
'I kept records, in the docks. I have French and English as well as Vietnamese: the languages are thought to be useful.'
'In custody for six years, considered by the authorities to be suspect because of association with Westerners, yet you were allowed employment by that suspicious government!' Erickson's voice was hard now, a definite cross examination.
Ninh shifted, blinking, and said, 'I told you, the languages were thought to be useful.'
'No one else in all Vietnam had French and English and Vietnamese! Surely those are the three most common tongues of the country?'
'Yes,' said Ninh. 'I suppose they are. But that was the job I was allocated when I was allowed to return from Hanoi.'
'You applied for it or you were allocated?'
'I was allocated.'
'Could you have refused?'
Ninh shook his head, bemused at the question. 'No,' he said. 'You cannot refuse.'
'Ho Chi Minh City,' said the lawyer.
'I am sorry, sir?' frowned Ninh.
'You refer to it as Ho Chi Minh City, not Saigon.'
'It is now called Ho Chi Minh City,' said Ninh.
Erickson nodded. 'Something else that cannot be refused?'
'That is the name,' insisted Ninh, doggedly.
'How are you released?' demanded Erickson, in a sudden switch.
'I do not understand the question,' said the Vietnamese.
'What is the procedure for re-education?' expanded Erickson. 'Are you committed for a definite period of time?'
Ninh shook his head. 'Just to re-education. It is up to the authorities to decide the period of time, when they are satisfied with your conformity.'
The questioning lawyer smiled, happily. 'Conformity,' he said. 'An excellent word. It took six years for the authorities to determine that you were sufficiently conforming?'
'Yes,' said Ninh.
'How?' said Erickson, shortly.

Ninh shook his head. 'How?'

'What was the moment, after six years, that made them decide you had conformed to a new doctrine?'

'There are appearances before examining committees. Questions. You have to satisfy the questions.'

Erickson said, 'After six years you satisfied those questions?'

'Apparently so.'

Erickson leaned across his table, towards the Vietnamese and said, 'Is that all?'

Ninh shifted and said, 'They were satisfied.'

'That wasn't my question,' persisted Erickson. 'What else is necessary to satisfy them that your attitudes are sufficiently changed.'

'You have to have a thorough knowledge of socialist doctrine,' said Ninh.

'Isn't it also necessary to avow that you *are* a socialist: that you've *become* a communist?'

'Sometimes,' said Ninh.

'What about your time, Mr Ninh? Were you asked to state that you now recognised your earlier mistakes and are a believing communist?'

The enquiry room was absolutely quiet, all the concentration upon the hunched Vietnamese. 'Yes,' he said quietly, head bowed. He brought it up, in weak defiance and said, 'But I . . .' but Erickson interrupted, unwilling to allow any escape. 'You did what you were told by the authorities!' he said. 'You learned your rote and you adopted the creed and you were allocated a job! You conformed to everything they wanted!'

'I had to get out of the camps,' said Ninh, desperately.

'So you lied?'

'Yes,' nodded Ninh. 'I lied.'

'You knew what a dog-tag looked like? Had seen them many times?'

'Yes,' said Ninh, curiously.

'You knew the names of the men involved on the Chau Phu mission, because you worked for Edward Hawkins?'

'Yes.'

'In the camps you heard many stories of Americans still imprisoned?'

The questions were coming like punches and Ninh was

physically wincing to them. Katzback had been intentionally mild, to lull Ninh, Hawkins thought.

'Yes, many stories,' confirmed Ninh.

'But never saw any?'

'No.'

'Would you have done anything, anything at all, to avoid going back to a camp?'

'Oh yes,' said Ninh fervently. 'Anything.'

'A camp you feared you might be returned to, after being contacted by the son of Edward Hawkins?'

Ninh nodded, uncertainly.

'Lied?' pressed Erickson. 'Would you have lied to have avoided going back?'

'I fled, instead,' pointed out Ninh, innocently.

'I did not mean lie to the Vietnamese authorities, Mr Ninh,' said Erickson, establishing his point. 'I meant create a story you knew would make you acceptable to the West. And lie to us.'

Hawkins felt Jones lean towards him and responded to the whispered approach. 'I hope you're still convinced,' hissed the assistant editor. 'I'm not.'

Surprisingly the investigation did not continue with Nicole Tiné, as Hawkins expected, but instead with the group he'd already guessed were CIA employees and who were confirmed as such when they were called. The first identified himself as Henry Wilder, a graphologist attached to the Agency's Technical Division. To the assembled lawyers he handed copies of a report and said that a week earlier he had been given examples of a signature for comparison. The name had been Nguyen van Ninh. It was inscribed upon a document he had been able to establish, from paper and ink samples, as that issued by the Military Assistance Command, Vietnam, to correspondents during the American presence there. The comparison was with a separate signature upon paper letter-headed with the name of a British freighter, the *Henry Clair*.

'In my opinion,' said the graphologist, 'The signatures could have been made by the same person.'

'*Could*?' seized Erickson at once, recognising the qualification.

'Yes, sir.'

'Not were?' persisted the Presidential lawyer.
'No, sir.'
'What you are telling us then, Mr Wilder, is that you found it impossible to reach a positive conclusion that the signatures upon the accreditation document and the letter-headed paper were made by one and the same man?'
'I found it impossible,' said Wilder.
Beside him Hawkins felt Harry Jones shift, uncomfortably, and then saw him lean sideways towards the diplomat from the British embassy.
'How long have you been employed by the CIA?' asked Katzback, taking up the other lawyer's point.
'Fifteen years, sir.'
'You are the senior graphologist of the department.'
'I am.'
'Are such difficulties common?'
'Not uncommon,' said the expert.
'On average,' came in Erickson, 'How often are you able to make a positive identification?'
Wilder considered the question and said, 'Between sixty and seventy percent.'
'How far short of that average does the signature of Nguyen van Ninh fall?' said Erickson.
There was another hesitation from the expert. 'I would say fifty-fifty.'
The next witness from the Technical Division identified himself as a forensic scientist named Walter Abler. Like the previous expert he handed the lawyers copies of a report before commencing his evidence. His tests had been upon metal discs inscribed with the names Forest, Bartel and McCloud. The metal was an alloy of American manufacture of the type used for military ID and the lettering appeared consistent with the engraving of those discs. They were extremely dirty and worn and upon those in the name of Forest and McCloud there was damaging to the edge, as if they had been used upon other metal, in the manner of undoing screws or removing caps from bottles or tins.
Once more it was Erickson who showed the scepticism. 'Are you aware of the great amount of American manufactured metal left behind in Vietnam after the American withdrawal?'

'Yes, sir,' said Abler.

'Would it have been possible for these IDs, providing a person was sufficiently acquainted with them, to have been manufactured and engraved in Vietnam?'

Abler frowned. 'It would have required a knowledge of the type of alloy, sir.'

'But not impossible.'

'Not impossible,' agreed the forensic scientist. 'But I would consider it extremely unlikely.'

'Well!' Hawkins said to Jones, in a whispered demand.

'Not enough,' insisted the other man.

The special Task Force assembled immediately after the enquiry, in the Oval Office again, everyone remaining quiet while William Erickson reported on the first day of the enquiry. At the end the President said 'OK. What's your opinion?'

'Doubtful of the signature: I think we've got to take the point that the forgery of the IDs would be difficult, though,' said the counsel.

'So where does that leave us?' asked Harriman.

'Little more than before,' offered Keys, the Secretary of State. 'Still without any corroboration.'

'There is one thing,' came in the CIA Director. 'I asked the French for help, as was suggested.'

'And?' said Harriman.

'There was a legionnaire in Vietnam from 1951 through 1953 named Marcel Tiné. Decorated for valour, on two occasions. There's no record in any French military archive of Marcel Tiné ever having been married in Vietnam. His wife's name is Claude. She's native-born French and she's still alive, in Marseilles. He died in 1962. There were two children, both boys.'

'What the hell does that mean?' demanded Harriman, with growing exasperation.

'There's something else,' said the Agency Director. 'There's no record of anyone named Nicole Tiné ever having worked at the French embassy in Saigon during the period the woman claims to have been there.'

'Nothing at all?'

'I'm taking it back as far as I can,' assured Snow. 'There was a Nicole Dulac, a Nicole Farkas and a Nicole Vingh. Paris think there might be file pictures but they can't guarantee it.'

'Shit!' said the President.

Jones insisted upon a private conversation with the embassy diplomat directly after the hearing and then rode unspeaking back to Washington, his right knee vibrating constantly under his hand, an indication of annoyance. No one else spoke, either. At Maryland Avenue Jones gestured the refugee couple abruptly into the kitchen and said to Hawkins, 'I want to speak to you alone.'

When they reached the study Hawkins angrily confronted the man and said, 'What the hell do you think you're doing, in my house! You're treating them like shit!'

'Maybe that's the way they deserve to be treated.'

'You're panicking,' accused Hawkins.

'Do you know what Smale said, afterwards! That he wasn't convinced: that there seemed some irreconcilable difficulty in the account. Can you imagine the sort of report he's making to Whitehall, from the embassy right now!'

'If he's pre-judging things on what he heard today, like you are, then he's a fool,' said Hawkins, carelessly. 'There's more evidence to be heard yet, from Nicole.'

'Crap, like today,' dismissed the man.

'The ID wasn't disproved.'

'What *was* proved!' said Jones. 'I'm going to warn the office.'

'Warn the office about what!'

'An impending disaster,' said Jones. 'This is going to blow up right in our faces. I've got a responsibility to the newspaper to tell them how today's gone and I've got a responsibility to warn them that it isn't looking good: isn't looking good at all. I'm not going to be associated with another Fleet Street hoax: there've been too damned many of those.'

'Make it your personal opinion, not mine,' said Hawkins.

'I always intended to,' said Jones, heavily. 'Believe me, I always intended to.'

Hawkins wanted the bastard to go, so he could have a drink.

Chapter Nineteen

Nicole Tiné pulled her hair back from her face and knotted it tightly into a bun, a style Hawkins hadn't seen her use before. She avoided any make-up and wore a formally tailored, grey check suit from the hurried purchase at Garfinkels in the time they had between their arrival in Washington and the beginning of the enquiry. It took the same form as before, Katzback gently leading. Nicole responded to the politeness, demure and small but speaking in a firm, even voice, making a much better initial impression than had Ninh: Hawkins hoped the concluding impression would be the same. To Katzback's considerate prompting Nicole said that she had fled Saigon before the actual moment of communist occupation, going after Edward Hawkins' departure to Dalat to be with her mother. She remained there for six months, almost seven, gradually coming to believe she was going to be ignored by the authorities. It was in November, the first week, when the cadres began their identity checks, working from lists that had apparently been created in the capital. There were twenty arrested with her, although she was the only one of mixed parentage. There had been a further three months in a detention camp, awaiting a hearing, and then the session itself. By remaining in the country she had surrendered her right to French nationality, she had been told; she was Vietnamese and had been corrupted by her association with Westerners. Her first camp was in the north, beyond Hanoi. She'd remained there the longest, almost four years. They'd had to work in the fields but also attend the lectures. She'd also been used for some translation work, propaganda material, Vietnamese into French and French into Vietnamese. After Hanoi there was a move slightly south, to

Huê. Here her use of French had been utilised more fully. The Vietnamese appeared to have a great deal of written French material, some of which she recognised from her earlier employment as being from French embassy or consular sources. Her function was to translate them into Vietnamese. She had been there two years after which came the transfer to Can Tho. It was here she encountered the Americans for the first time, although like Ninh she had heard rumours of prisoners during her movement throughout the country.

'This is important,' said Katzback, coming forward over the table. 'We want as detailed a description of this camp as you can give us.'

'It is quite large,' said Nicole, head bent in recollection. 'I would suppose two, maybe three hundred people. A large staff, too. Guards and political cadres. Probably fifty. There is the river, which is the supply route. Two jetties. It is constructed . . .' She hesitated, searching for the word, rotating her hand in a circular motion in front of her. 'It is round,' she said. 'In the very centre there is the political compound, where the Americans are. The hall is there also, where the lectures are given. And the other building, where the indoctrination is conducted. This is the most heavily guarded, with a proper perimeter fence, wire and electrical sensors. In the next circle there are barracks for Vietnamese political prisoners. It is much bigger, of course, and they have to go into the central meeting area for the meetings. There is a third circle and here the common criminals are held.'

'Criminals?' pressed Katzback.

Nicole nodded. 'It is the system. Apart from the recognised guards, the criminals act as additional protection. If anyone escapes through their ring, the criminals are punished. If they detect someone, they are rewarded with a lessening of their sentence.'

Katzback frowned to the other two men on either side and said 'You made a division, between the lecture centre and the indoctrination hall. Aren't they the same?'

Nicole shook her head, positively. 'The lectures are for everyone: indoctrination is individual, by psychiatrists and psychologists. That is where the Americans are treated.'

'Tell me about this treatment,' said Katzback.

'I was at the camp two years,' said Nicole. 'There are three hours in the morning, four in the afternoon. Sometimes the Americans are together, sometimes separately.'

'Are they ill?'

The woman nodded. 'They have sores. And their teeth are crooked, I think it is because of the diet. They are fed only twice a day, the first time at noon and again in the evening. It is usually soup. As well as attending the lectures, they have to work in the fields, cultivating for the camp staff. If they are caught trying to take any of the vegetables for themselves they are punished. There is a small wooden building, alongside the indoctrination hut, in which they are locked. It is very small, like a cage: it is impossible to stand up. Anyone inside has to crouch, like this . . .' She made the movement of pulling her knees beneath her chin. 'And they are very confused,' she said.

'Confused?' demanded Erickson at once.

'There are drugs used upon them. There is a man called Page, a black man. He is very big, still strong despite his suffering. Some of the guards seem frightened of him. He is allowed to stay in his cot more than the others. He has no mind, not any more. He laughs sometimes: sometimes he just lies and stares at nothing. He cannot care for himself . . .' she frowned '. . . the toilet I mean. That was one of my duties, to clean him. The others were frightened of infection: the guards, too.'

'What about the others?'

'The man called Forest is still the strongest, in his mind. But he's very weak, physically. Sometimes he has to be helped by the others, particularly Bartel, when they go out into the fields. Both Bartel and McCloud try to do his work for him. Just before I was released Bartel began crying a great deal. McCloud was growing quiet, sometimes like the black man, Page.'

'What contact did you have with them?'

'Most days,' she said, 'I had to clean Page, as I have said. The indoctrination is in English but the Vietnamese who give it live outside the camp. They don't involve themselves in the running and the administration. I had the language so I acted as interpreter.'

'Tell me about the tags?' said Katzback.

'I told the man Forest that I was being released: being sent to a resettlement camp before being free. He made Bartel and

McCloud take their discs off and gave them to me, along with his. He said I was to go to a Western embassy. That the discs would mean something and that I was to get help. That if I did I would be helped too.'

Throughout the session Erickson had written steadily. Now he made a gesture, to the chairman, who nodded and Hawkins realised it was a rehearsed procedure because Erickson immediately attacked. 'You had been imprisoned, in re-education camps for several years?'

'Yes,' said Nicole.

'In conditions of great deprivation: suffering?'

'Yes.'

'Americans caught stealing were put into a small cage, in solitary confinement?'

'Yes.'

'What happened to others caught infringing regulations?'

'They were punished, too.'

'How?'

'Sometimes beaten. Sometimes the food ration was cut down.'

'What about further imprisonment?'

'That too.'

'Madame Tiné, having been imprisoned for several years in conditions of great deprivation, why did you agree to carry out of a camp identity discs of Americans, knowing as you surely did that if they were discovered you would be punished, probably with the severity of a further sentence!'

Nicole looked at once towards Ninh, as if for help, then to Hawkins. She said, 'I thought it would enable me to get out too.'

'From Vietnam?' persisted Erickson.

'Yes,' said Nicole, quiet-voiced.

'Your father was French?'

'Yes.'

'Before your association with Edward Hawkins, you worked at the French embassy?'

'Yes.'

'You had the right to French nationality. And you would have known the system for leaving Vietnam for France. Why didn't you go before the end of the war?'

'I would not leave my mother.'

'But now you didn't mind?'

Nicole stared down into her lap. 'When I got to Dalat I discovered she had died, two years before.'

'*Did* you work at the French embassy?' said Erickson, prepared from the previous night's session with the Task Force.

'Yes.'

'I will repeat the question. Did you work at the French embassy in Saigon?'

'Yes!'

Beside him Hawkins was aware of Jones moving. The man said to him, 'Christ am I glad I sent the warning.'

'Madame,' said Erickson, pointedly avoiding the surname. 'We have checked with the French authorities, throughout the period we understood from you that you were employed. Paris say there was no one named Nicole Tiné at their embassy in Saigon.'

From Jones alongside there was a sigh, either of resignation, or impatience, Hawkins couldn't decide which: didn't want to decide which. Fuck, he thought; oh fuck!

'I worked at the French embassy!' insisted the woman.

'So the French records are wrong?'

The woman's composure was going now. Her face was flushed, breaking up, and she was tugging apprehensively at some gloves coiled through her handbag strap. 'No,' she said, so quietly it was difficult to hear. 'No, they are not wrong.'

It was Katzback who asked the question, reverting to kindness. 'We don't understand,' he said.

'Vingh,' said the woman. 'My name is Nicole Vingh. My father *was* French. His name was Marcel Tiné and he was with the French army . . .' She was tugging almost irritably at the gloves now. 'They were never married,' she took up. 'That's why I had no French residential rights . . . no rights at all. Why my mother and I couldn't go to France.'

It was very quiet in the room. The three lawyers moved together, in a huddled conversation. It was Katzback who continued the questioning. 'Why did you lie about the name? Why didn't you say Vingh?'

She looked pleadingly towards Hawkins, crying openly, and the explanation was disjointed and broken. 'I always used Tiné, except when there were official papers and I couldn't, because I would have been found out . . . there was respect, to be French. I look European . . . to have a name like Vingh marked me illegitimate. Edward knew me as Nicole Tiné . . . I thought his son would too . . .' She made a helpless movement. 'It seemed such a little lie . . . it became one I couldn't escape from . . . never . . .' She took a handkerchief from the bag in her lap and blew her nose, a determined effort for control.

'What a lot of crap!' said Jones, from beside him.

'We know about Nicole Vingh,' said Katzback. 'That someone of that name worked in Saigon, at the embassy.'

Hawkins looked sideways at the man he despised and said, 'Crap?'

Erickson took up the questioning again. 'We heard from Mr Ninh of the requirements to satisfy your captors you fully adopted the Communist regime, before release?'

'Yes,' she said.

'Did you satisfy them?'

'I must have done. I would not have been released if I hadn't.'

'Did you take an oath to that effect? Give an undertaking?'

'Yes.'

'You told them you embraced the Communist doctrine?'

'It was the only way to get out of the camp.' Her face was flushed at the awareness that the pressure hadn't relaxed.

'So you are a Communist?' demanded Erickson.

'No!' said Nicole. 'It was the way out: the only way.'

'Which took you all those years to discover?'

'No,' she said again, desperately. 'Everyone says it; of course they do. So the authorities disregard it for a long time for what it is, a trick to get out. It is only when they *believe* that you get released.'

'Where do you think the Vietnamese got the documents you were required to translate, at the camp at Huê?'

'I do not know,' she said.

'Some of them were embassy documents: official?'

'Yes.'

'Stolen then? Or looted?'

'I do not know,' she repeated.

'Were some of them marked Confidential.'

Her head was in her lap now, her voice quieter still. 'Yes,' she said.

'Some marked Secret?'

'Some,' came another almost inaudible admission.

'You were part of a spying operation then?'

Her head came up, her face contorted. She bit against fresh tears and Hawkins saw that her hands were white, clutching at the crumpled, soiled gloves. 'I was a prisoner, in a camp,' she said. 'I didn't have any choice: I had to do what I was told.'

'Like you were told to act as an interpreter for the Americans at Can Tho?'

'Yes,' she said, wearily.

'Their orders, the instructions telling them what to do, came through you?'

'Not all the time.'

'How much of the time.'

'If there were orders that were unusual, something they weren't used to, then I had to explain.'

'What about the indoctrination . . . the changing of their minds?' said Erickson.

'No!'

'What about punishments? Who explained the punishments to them, when they had broken some regulations?'

She shook her head, not answering.

'Did you ever explain or inform an American he was going to be punished?'

'I had to do what I was told,' insisted Nicole. She appeared dulled and lethargic, almost as if she were disinterested. She made an effort for recovery, bringing her head up to confront Erickson. 'I've told the truth,' she said. 'I know I told a lie and I know it makes you doubt me but I've told the truth . . . the absolute truth. There *are* Americans. Please believe me!'

It didn't seem that Erickson did. He said, 'How did you get those identity discs?'

'I have told you, from the man Forest.'

'Weren't they made, by Ninh? Weren't they concocted, like this whole story, in the hope of gaining permanent residence in this or any other Western country!'

Nicole's attitude now was one of almost complete bewilderment. 'NO!' she said.

'Tell me about your release,' persisted Erickson, relentlessly. 'The procedure, I mean. Did they just open a gate and turn you loose?'

'There was transport, back to Ho Chi Minh City: that was the point from which I was sentenced.'

Erickson nodded and said, 'But before that? Did they just open the gates?'

Her bewilderment remained. 'Well . . . yes . . .'

'What about last minute procedures? Weren't you allowed a change of clothes . . . a bath . . .?'

'There were clothes, a suit made of some canvas type material. And a shower.'

'What about a search? Were you subjected to any sort of search?'

'Not really,' she said.

'Escorted?'

'Yes.'

'Tell me, if you were issued with fresh clothes and those you wore were taken from you, and if you were escorted at all times, how did you manage to hide the identity discs?'

'That was not difficult, for a woman,' she said.

Momentarily Erickson failed to understand. He said, 'How . . .' and then stopped and Nicole said, 'I carried them from the camp internally, like a sanitary tampon.'

Erickson flushed, for the first time during his interrogation off-balanced by a reply from her. Trying to cover his discomfort he said, 'You describe your imprisonment as one of deprivation and hardship?'

'Yes.'

Erickson looked out into the room at Ninh, then back to her. 'You don't appear to me to look like someone who's suffered hardship and deprivation. You appear to me to be an extremely attractive woman whom it's difficult to believe underwent nearly seven years of imprisonment.'

Nicole straightened in her seat, gazing across at the man defiantly. 'That deprivation and hardship was at the beginning.'

'Before you began cooperation,' seized Erickson at once. 'Before you became the translator, someone who was converted!'

'I did not get the favours that way, not by that sort of cooperation.'

'Explain that reply,' insisted Erickson.

'I have the appearance of a Westerner,' she reminded him,

her voice level and strong now. 'And I am a woman. Which made me attractive to the men who ran the camps: every one I was in. I learned that in Hanoi, and it was the same everywhere I went. To avoid working in the fields and to avoid punishment and to get more food than anyone else . . . to survive . . . I slept with anyone who wanted to make me their whore . . .'

There was a shift of discomfort throughout the room.

'. . . there were a great many,' continued Nicole. 'That was the worst hardship of all . . .'

They arranged the meeting in a series of guarded telephone calls, twice changing the rendezvous because Eleanor was nervous of the choice. Hawkins became irritated and embarrassed at the theatrical subterfuge of it all and when they finally met, in the bar of a Holiday Inn in Kensington, way beyond the Capital Beltway, it was obvious she felt the same way.

'This is like a play that Neil Simon wished he hadn't written,' she said. They had the darkest booth they could find and she sat with her back to the door.

'Shouldn't you take your sunglasses off?'

'This is ridiculous,' she said.

'Yes,' he said. 'I suppose it is.'

'Someone's moved in, haven't they?' she said accusingly, misunderstanding.

'Not that way.'

'What other way is there?'

'It's not a woman.'

'Who then?'

'I can't tell you.'

'You're not making sense.'

'I've missed you,' he said.

'Can I have another drink?'

Hawkins indicated with his glass to the distant barman and repeated 'I said I missed you.'

'Which you prove by saying I can't come to your house any more and make us go through this goddamned stupid routine.' She stopped when the drinks arrived, turning away with some feigned preoccupation with her handbag. After the man left she looked back to Hawkins and said, 'You want to end it?'

'I said it's not a woman.' She already knew of the existence of Ninh and Nicole: he'd told her after Vietnam.

'And I said you weren't making sense.'

'It's work . . . people involved with work.'

'How long are they going to be with you?'

'I don't know.'

'Why the hell are you being so awkward!' she said, drinking heavily from the glass. 'A day! a week! a month! How long?'

'Maybe a long time: maybe months.' This time it was Hawkins who signalled for refills.

'We've been honest with each other,' said Eleanor. 'I wasn't at first but I admitted it. So now I'm going to admit something else. I told you why it was, why I came to the house and all that stuff. But I didn't intend to let it get out of control. It was supposed to be us, the little people who fall backwards into whisky glasses: who knew the signs and could maybe give each other a little comfort. But that's all. I wasn't supposed to fall in love with you because falling in love with you is preposterous and impossible and can't be . . .' She stopped when the man arrived with the drinks and this time she didn't try to turn away. When he'd gone she went on: 'And now I know it isn't the same with you, despite what you said before . . .' She dropped her head, as if it weighed heavily, looking into her glass. 'Another fuck-up,' she muttered. 'Like all the rest.'

Hawkins felt warm, from hearing her finally say it and from the drinks, the beginning of the familiar numbness in his cheeks. He could trust her, if she loved him; he didn't have to tell Eleanor what they claimed, after all.

'I love you,' he said. And then he told her. He was extremely careful, saying nothing of the supposed survival of Forest and Bartel or McCloud, just about the refugees' escape and she stared at him while he talked, blinking and with her lips moving, finally forming into a smile.

'But that's amazing!' she said. 'Absolutely amazing!'

'Yes,' he said. He wondered what Jones and Smale had already reported back to London.

'But why didn't you tell me! Why make me go through hell and imagine you were going to dump me?' This time she waved her glass around the booth partition, as a signal.

'It's . . .' groped Hawkins, trying for an explanation that she would understand and would not involve him in difficulties. 'It's important no one knows . . .' he said. 'Not even John.'

She laughed, a dismissive sound. 'That's hardly likely, with the sort of communication that exists through us.'

'It's important,' he emphasised. 'Don't tell anyone. Promise me?'

'If it's important of course I promise.'

'It is,' he insisted.

The man arrived with yet more drinks and Eleanor squinted up at him and said, 'Do you realise you're serving the wife of the future President of the United States?'

'Whatever you say ma'am,' said the waiter, who'd seen it all before and was professionally blind.

'Don't be silly, Eleanor,' said Hawkins awkwardly, after the man had left.

'Don't worry, darling: I won't embarrass you,' she said. She grinned crookedly at him and added, 'Know what I'd like?'

'What?'

'For you to go and get a room.' Seeing the look on his face, she said: 'But I'm not going to let you. I'm not that drunk. Or stupid.'

'Shit!' he said vehemently. 'Why doesn't anything work!'

'Because nothing works,' she said, with inebriated logicality.

When Hawkins got home there was a repeated message waiting on his answering machine, instructing him to call London at once. When he tried, the London switchboard routed the call through to Wilsher's home and from the reply Hawkins realised the editor had been asleep.

'I've been waiting for hours,' said Wilsher, irritated.

'I'm sorry,' said Hawkins. 'I had an appointment.'

'We don't like what we're hearing back here,' said Wilsher. 'Doondale wants a full report; your side of the story. So do I. The impression here is that we're involved in something potentially embarrassing.'

'Jesus!' said the President. 'Holy Jesus!'

'That's what I thought,' said the CIA Director.

'Jesus!' said Harriman again, for once in his political life unable to think of anything else to say and for a politician of Harriman's experience that was something that had never happened before.

'Would you like the experts to make a presentation?' asked Snow.

'You're damned right I would,' said Harriman, recovering.

Chapter Twenty

An effort was made to use the period it took to move the visual presentation equipment into the Oval Office to assemble the already established Task Force and contact others to be included, upon the President's instructions, but it was late evening and people were not readily available. The Secretary of State, for instance, was at an official welcoming banquet for the French Foreign Minister and could not immediately leave, Richard Godsell, the Defence Secretary, was attending a violin recital by Yehudi Menuhin at the Kennedy Centre and Ben Jordan was at a farewell party for a retiring Agency officer at Chevy Chase and it took two hours to locate him. Jordan was one of the additions to the group, along with Pearlman. So were Katzback and Paulson, the two lawyers comprising the enquiry panel with the Presidential counsel.

The delays meant it was past midnight before the meeting began in an office crowded because the area in front of the fireplace was occupied by several covered easels, greatly reducing the space. Chairs had been arranged in front of the easels but the President, who'd had everything explained to him by the technical expert before the meeting, remained at his desk.

Harriman nodded to the CIA Director, who remained alongside the easels instead of sitting on the arranged chairs, and said, 'It's late: let's get going.'

Next to Milton Snow was a slim, straying-haired man who had not had time to change from his workaday windcheater and the previous day's shirt and who was clearly uncomfortable at the way he was dressed and at being in the President's office anyway, the focus of attention.

'This is Alan Dobble,' introduced the Director. 'He's a

senior photo-analyst in the Technical Division of the Agency. He'll explain what's behind these covers but for those of you who weren't involved from the beginning let me explain they come from a satellite moved directly over the Can Tho area . . .' To the nervous man he said, 'OK.'

Dobble coughed and lifted the covers from the three prepared blackboards. There was a strain forward of interest from the seated group at the selection of photographs which were pinned to them. From the first easel Dobble took up a sharp pointer and said, uneven-voiced, 'These are photographs which were relayed back over a period from that satellite, yesterday . . .'

He moved back, to the first board, and continued, 'They are numbered, in order of their being received. You will note . . .' he tapped schoolmasterishly with the baton '. . . that the early photographs are very much more indistinct than those numbered later . . .' There was another gesture with the pointer, to the second and third boards. 'The reason for this is that having thought we recognised something, we asked for more detailed repositioning for positive confirmation.' The man coughed again, confident with his subject and losing the uncertainty. 'Even from these early frames, however, you will be able to see definitely that there is an installation, constructed in what appears to be gradually expanding concentric circles . . .'

There was further movement from among the group, from people who had attended the enquiry and heard the description from Ninh and Nicole.

Dobble continued: 'There are two sets of the later photographs, one clean, as they were received, the other tagged for the benefit of clarity for what I am about to explain. The arrowed figures are human. You will be able to see from frames four and eight what appears to be some form of regimentation, such as might be expected in a camp. The frames to which I would like you to pay particular attention are nine, eleven and thirteen. You will be able to see, because they are quite clear, the indication of those figures creating shadows: according to the timing which is superimposed, frame nine was taken at 13.15, eleven at 14.06 and thirteen at 15.40. Those timings are important, to show exposure after midday, when shadows would be formed.'

Dobble allowed a break, coming to a point in the evidence

he wanted the assembled group to recognise. He said, 'Great accuracy is possible, in photo-reconnaissance. From those shadows, the height of the people casting them can be calculated. In the opinion of our experts, the people photographed in these frames, a total of four in all, are of Caucasian, not Asian stature. Westerners, in fact.'

Dobble turned back out into the room. 'I know, of course, of the reason for this satellite reconnaissance. The estimated height of the Caucasians photographed in these pictures is between five-ten and six-two. According to their military records, Colonel Forest was six-one, McCloud was five-nine and Bartel five-eleven.'

There was open muttering, a swell of noise from the watching men.

'Any questions?' invited the CIA Director.

'Yes,' said Erickson, at once. 'You've described the camp as being arranged in concentric circles. Can you tell me what exists in the very middle of those circles?'

'Three buildings,' replied the photo-analyst at once. 'Two quite close, the other distanced.'

'What about the next circle?' asked Paulson, the other lawyer.

'More buildings,' said Dobble. 'Constructed in a regular arrangement.'

'Like barracks?' said Erickson.

'They could be barracks,' agreed the man.

'How many circles, in total?' said Katzback, entering the questioning.

'Three,' said Dobble, immediate again.

'Is there something about the geographical positioning of the camp with which you can help us?' said Erickson.

'Geographical?' queried the young man.

'Yes,' said Erickson, refusing to indicate the answer he hoped for.

Dobble went back to his photographs, then smiled back out into the room. 'I'm sorry,' he said. 'The installation adjoins a river: from our examination of the bank contour it would appear there has been some widening, at the actual point of the installation to permit loading and unloading. There seem to be sheds at the bank.'

The three panel lawyers exchanged glances and nods and

Paulson said, 'Anything else particular about the river and the installation?'

There was further hesitation from the expert as he tried to understand the question. Then he said, 'Jetties. There are two jetties, about twenty-five yards apart.' He gestured with the pointer. 'The sheds are where the jetties abut the bank.'

Speaking generally to the room Erickson said, 'That concurs, in every single detail, with the description of the camp given by the woman.'

'I would like you to consider this,' interrupted Milton Snow, taking a photograph from a case propped against the first easel. 'I was going to have it introduced at tomorrow's hearing but there doesn't seem any point in holding it back, in the circumstances.' He held up the photograph, then offered it to those comprising the enquiry.

'It's her,' identified Katzback. 'Where did you get it?'

'French Foreign Ministry,' said the CIA chief. 'It's a file shot of a woman who worked at their embassy in Saigon until 1973. Her name was Nicole Vingh.'

'Which is what she admitted to at the enquiry,' said Paulson.

To his own counsel the President said, 'OK, give me a breakdown of what we've got, positively.'

'A disputed signature of a man who described that installation pretty accurately, in its initial construction,' responded Erickson. 'Pretty definitely – definite enough for my satisfaction anyway – bone fide American dog-tags. The precise description from the woman of the camp, as I have already said. And positive photographic identification by her of photographs of Forest, Bartel and McCloud in a test when she was invited to make the wrong choice but didn't.'

To the other two lawyers Harriman said, 'Any dissent?'

'No,' responded Katzback and Paulson, almost together.

'What about her lying, over her name. I don't understand that?' said Harriman.

'People do lie, in apparently stupid ways,' said Paulson, the professional lawyer. 'I agree it appears ridiculous to us but I accepted her explanation, given her circumstances.'

Again the two other lawyers nodded in agreement.

'Any other views?' said the President.

'I go along with the evidence being convincing,' said Jordan, who wished he hadn't drunk beer at the party because now he wanted a lavatory. 'But I don't think we should lose sight of the fact that the Soviet Union now supports the regime in Vietnam. There are thousands of advisors in-country: Russians are Caucasian.'

'Good qualification,' agreed the President. 'Anyone else?'

'I think the evidence is too conclusive the other way,' said Defence Secretary Godsell, solemn-voiced. 'I think we've got the first confirmation after more than eight years of American servicemen still being held in captivity in Vietnam.'

Ironically the public expression of something they were all considering had a sobering effect in the room, subduing the rising, expectant excitement.

'I think so too,' said General Cornell Bell, the Army Chief of Staff, suitably grave.

'You've comprised the panel at Langley,' Harriman said to the lawyers. 'What's your feeling?'

There was a momentary exchange of looks between the three men, then nods and Erickson said, 'We think it's confirmation, too.'

'Anyone disagree?' said the President.

There was complete silence within the room, which Harriman allowed to continue for several moments. Then the man said, incisively. 'Neither do I. We go in to get those poor bastards out.'

There was a break, for Dobble to clear and take with him his display equipment for which everyone was grateful, because it provided more room and for which Jordan was even more grateful because it permitted him to use the bathroom. Harriman ordered sandwiches and drinks, beer and soda and coffee, which was served in the Oval office. People formed predictable divisions, the three lawyers together, the Secretary of State in conversation with Pearlman, Godsell and the Army Chief of Staff drawn together and Snow with the returned, relieved and non-drinking Jordan. Between them all moved the President, listening more than talking. After an hour and lavatory visits by most others Harriman said, 'OK. It's getting late. Let's continue,' and the groups split into the previous, more general assembly.

Harriman waited until they settled down and became quiet

and said, 'This is the concept. I want it quick, I want it to work – without any failure – and I want them out. Don't forget the absolute disaster of the Carter mission to free the hostages from Iran. A repetition of that I *don't* want. Apart from the Britons you all know about, we're the only people involved and that's the way it's going to stay because if there's a leak – the vaguest suspicion of what's happening and why it's happening – we've got every newspaper front page and television screen in the world. And guys we all believe to be Americans and who have suffered for the past eight years are going to suffer a hell of a lot more. That understood?'

And I won't get the avalanche re-election, he thought.

There were movements and sounds of assent from the other men.

To the Army Chief of Staff the President said, 'Looks like your show.'

General Bell nodded and stood, a tanned but gaunt-faced man, iron-grey hair cropped into a crew cut. On the little finger of his left hand he wore a college ring, ornate gold encrusted around a red stone, and as he talked he twisted it with his other hand, revolving it around and around the finger. 'The aerial reconnaissance will give us a positive co-ordinate,' he said. 'The river's an advantage, of course. The feasibility study, knowing the location of Can Tho, but not the camp, concentrated upon a waterborne entry. Against Can Tho – before we had the co-ordinate – it was estimated we could make a waterborne entry, an assault and unimpeded river return in thirty-six hours . . .' Bell coughed, then continued . . . 'With the sort of mobile, absolutely trained squad we are considering, we think the chances of getting *in* are good. That still puts Can Tho within a day, two days of entry. Obviously the return is the problem, irrespective of others to which I will come in a moment. Directly an open attack is made, the assumption must be made that the return is jeopardised.

'The intention is to despatch some of the entry group down river as a decoy: run maybe for two or three hours, until they're sure they've picked up pursuit and then go into the jungle. The sort of men I'm talking about are the absolute professionals in the Special Forces: their function will be to run hare, to attract the hounds . . .'

'Meanwhile?' prompted Harriman.

'Meanwhile the real assault group, with the rescued men, ignore the river. They cross overland to a down-river stashe point, pick up a secured boat and come out into the open sea for a rendezvous from the original departure craft.'

'What about the decoy group?' demanded the President. 'I don't see we're achieving a lot exchanging one group of Americans for another.'

'The decoy group will be two, possibly three men: highly trained, behind-the-lines incursion specialists. They remain for a week, maybe more, after we've cleared the area with the rescued prisoners. Then we make an announcement of the rescue, giving the impression that our stay-behind group haven't stayed behind after all. We'll have support vessels in the area, to arrange pick-up from radio contact, to be made at prearranged times and frequencies.'

'Jesus!' said the President. 'That's certainly the other extreme!'

'I don't think this should be looked at as anything other than a high risk mission,' said Bell. He looked briefly towards the CIA Director and added, 'This isn't a country in which we've got any agent structure: any friends we can call upon.'

'They'll be as exposed as hell,' said Harriman, as if the awareness was a sudden one. 'We'll be risking more Americans than we'll be getting out.'

'Yes,' agreed Bell, objectively. 'We will.'

'Any thoughts on this?' offered the President.

'Couldn't we *increase* the commitment, in an effort to guarantee success?' suggested Godsell. 'Why can't the incursion start as outlined but once there's positive, on-ground confirmation, the actual rescue be mounted by helicopters from some offshore carrier.'

'No!' said the President, positively. Everyone stared at him, caught by the force of the rejection. 'It is high risk,' said Harriman. 'So I've got to minimise that risk, as much as possible. The rescue group has got to be Special Forces because no one else has the expertise. Or ability. But I – we, America – have got to have some deniability.' He looked to the CIA Director. 'What cover can you give?'

Snow gave a vague movement of his shoulders. 'We can provide a front, sure,' he agreed. 'But if we've got provable American servicemen actually in-country, operating a mission, I don't see what protection a front gives us: almost the reverse, in fact.'

'What do you think?' said Harriman, to the Secretary of State.

Keys fidgeted in his seat, abruptly the point of attention. 'I think we've got to go in,' he said. 'I also think we're as exposed as a bitch in heat. And if it fouls up we're going to be in real trouble. We're going to have a further bunch of American servicemen in captivity which is going to be hell here at home and we're going to have Hanoi screaming invasion and we're going to have Moscow orchestrating an international and particularly Third World War propaganda campaign about aggression and violation of territory.'

'Right!' agreed Harriman. 'So we *must* have deniability, like I said.' He turned back to the Army chief and said, 'What sort of squad were you considering?'

'Small,' responded Bell at once. 'Not more than eight men –' he stopped, conscious of the President's frown. 'They'll have surprise – initially at least – and they'll be the best eight men the Special Forces have got. And believe me, Mr President, they're pretty *special*. Numbers aren't a factor . . .' Bell stopped again and said, 'And I've got a suggestion here. We've planned the squad, in the event of it being necessary. And picked the person most qualified to lead it . . .' He stopped again and Harriman frowned, tired because it was approaching three am and irritated at the hesitation. 'Well?' he demanded.

'The Special Forces have an officer who has more combat experience than any other comparable commander,' said General Bell. 'He served four terms in Vietnam, speaks the language and is acknowledged the best incursion and behind-the-lines survivor in the Force: he's actually created a textbook used in training. He's about to be promoted General . . .'

'Then he's our man,' interrupted the increasingly annoyed President.

'It's Colonel Elliott Blair,' announced the Army Chief.

The surprised silence stretched out into the room, everyone waiting for the President to be the first to react. His response was typically that of a caucus politician. He looked around the room and said: 'Well?'

'We were considering recalling him for the enquiry, Peterson and Patton as well,' said the presidential counsel.

'If I can make a further point,' said Bell. 'Our feasibility study included psychological assessments. There's been evidence at the enquiry of mental stress: the psychological feeling is that these guys are going to be pretty well shot, mentally. There might be an advantage in having someone in the rescue squad whom the majority would recognise and respond to at once.'

'We need positive deniability, right?' asked the President, rhetorically.

There were movements and sounds of agreement from around the room.

'Then how about this for a scenario? How about the rescue attempt being led by a squad of *ex*-Green Berets, led by the *ex*-officer who was involved in the mission in which they were captured? It's happened in 1983. It fouled up. But they were *deniable* and there wasn't any outcry.' Harriman looked triumphantly out at the Task Force.

'Excuse me, Sir,' said the Army Chief. 'But Blair is a serving officer about to be promoted.'

'Why can't he be invited to be a former serving officer, just for the duration of this mission? Why can't they all be? They can tender resignations which can be produced if things go wrong. If it goes right the resignations get forgotten and there's a Presidential greeting on the White House lawn.'

'I don't think that's legally viable,' protested Erickson.

'We're inviting soldiers to volunteer,' bulldozed Harriman, unwilling to sacrifice his idea. 'You telling me soldiers have never been invited to volunteer?'

'In uniform they've got protection,' said Erickson. 'Out of it they're terrorists.'

'Which will be explained to them.'

'Can I offer something on this?' said the CIA Director, bending back from a conversation with Jordan. 'The other survivor from the Chau Phu mission, Eric Patton, is extensively involved with government transportation contracting, a lot of it with the Agency. He's also got a front company used by the Agency, for the reemployment of staff.'

'What sort of transportation?' asked the President.

'Everything,' said Snow. 'Air, sea and road transport: specialised raft, too.'

Harriman smiled. 'Could he provide a boat?'

There was a further muttered conversation between the Director and Jordan and then Snow said, 'We'd have to check but we think it's highly probable.'

'Are there provable links between Patton and ourselves?' said Harriman.

'With the military, probably,' said Snow. 'Certainly not with the Agency.'

'How much better can this get!' said Harriman, in increasing triumph. '*Two* men engaged in an original, foul-up mission involved in getting their buddies out, in an entirely private enterprise of which we can deny any knowledge – *if* it goes wrong!'

'I'd still like to get further legal opinion on this,' persisted Erickson. 'And there were some questions I wanted to ask all the survivors.'

The President shook his head, impatient at the caution. 'The enquiry is suspended: for the moment it's irrelevant. And I sure as hell don't want Peterson brought in, not in any way . . .' To General Bell, Harriman said, 'I want Blair invited to volunteer: tell him it's the personal request of his Commander in Chief . . .' The President continued on, to the CIA Director. 'The same for Patton. They both got it wrong once: now let them get it right . . .'

'There's the British,' warned Pearlman. 'They will want to know why the enquiry isn't continuing: there was a lot of trouble trying to exclude them in the first place.'

'We're still going to need the help of the Vietnamese pair in physically recreating a model of the camp for the actual assault training,' endorsed General Bell. 'They seem extremely dependent upon Hawkins.'

'We give them everything they want,' decided the President. 'We tell the British through their embassy here and we let Hawkins know he can personally travel on the mother-craft from which the actual assault will be made . . .' He paused, looking at Jordan. 'You'll go too. That way we'll have everything nicely in one basket, a civilian volunteer assault squad and the writer son of the journalist involved, well under wraps and unable to do anything until we give him permission.'

As the group broke up and began to leave the room Pearlman said quietly to the Secretary of State, 'I think I've just sat through the most cynical example of political manoeuvring ever.'

Willard Keys shook his head and said, 'You've just seen Washington politics at its best.'

The summons from the British embassy said it was important so Harry Jones insisted on going too, which made a wonderful moment for Hawkins when the ambassador made the announcement. General-designate Elliott Blair was told to abandon the review of entrant grades for the previous month and fly immediately to Washington for a personal meeting with the Army Chief of Staff. And Ben Jordan, with previous personal contact as Director of the clandestine division of the CIA, flew up to New York to interrupt Eric Patton just as he was about to address a specially convened meeting of his senior management to record the best year's trading profit after tax since the incorporation of the company.

'It seems you reached the right conclusions from the start,' congratulated Sir Neville Wilkinson.

'I'm glad I didn't rush to any premature conclusions,' said Hawkins.

'You've my personal assurance that if you agree and it fails neither you nor any of the squad who might volunteer with you will suffer, either from your military record or pensions,' over-committed General Bell. 'And if it succeeds your country will honour you.'

After a hesitation that was to be expected, for what he was being asked to do, Blair said, 'I understand. And of course I'll do it.'

'What do you say?' asked Jordan, after the explanation.

'Dead,' said Patton, head moving slowly in disbelief. 'They're dead . . .' Then, 'Oh God!'

Chapter Twenty-One

A few minutes drive from the colonially restored and maintained town of Williamsburg, in Virginia, there is a 10,000 acre installation officially named Camp Peary and unofficially called 'The Farm'. Its gate notice describes it as 'Armed Forces Experimental Training Activity, Department of Defence'. It is, in fact, a complex entirely utilised by the CIA, specifically for clandestine instruction and it was here the intensive training was carried out for the incursion into Can Tho.

Nguyen Ninh, who had been engaged in its construction, and Nicole Tiné, who had actually lived there, cooperated first with architects and then with builders of a mock-up of the Vietnamese re-education facility. To recreate as accurately as possible the water-served prison the incursion squad would encounter, it was constructed from prefabricated, erected-overnight materials along the banks of York River, which runs through Camp Peary.

From the Panama exercise in which his squad had been awarded top marking Blair knew the ten best men he wanted and when the mission was explained everyone accepted without a moment's hesitation the conditions that had been called upon from their Commander during the Washington briefing, as Blair had expected they would. During the design and construction of the Vietnamese camp, he took them out on to the river, rehearsing the rubber boat entry they would make from one of Patton's CIA designed and operated surveillance freighters which had been operating anyway in the Yellow Sea and Korea Bay and was already re-routed for the pick-up in the Philippines.

When the replica was built Nicole was further involved,

debriefed and debriefed over a full day and the following morning by a demanding Blair, wanting to know guard routines, manning levels, duty changes, roster strength, armaments and protection devices. Personnel at Camp Peary are accorded the highest classification and are accustomed to being involved in exercises the purpose of which they have no knowledge. Three platoons of such men were deputed as Vietnamese guards and a further two platoons as the criminals forming the outer ring of the encampment through which the Green Berets might conceivably have to fight. They made night and day assaults which they honed to stop-watch perfection. With a positive co-ordinate and able from it to calculate the precise distance from the river mouth a timed distance was run, at slow, avoidance speeds and flat-out, detected-and-running pace.

Blair was a man who was never militarily satisfied and it showed during the final briefing.

'Good, as much as any exercise can be,' he said. 'Don't forget we're working on information that is months old. Everything the woman told us could have been changed by the time we get there.'

'I don't think it's going to work, trying to bring those guys back overland,' said Wilbur Manson, the tobacco-chewing, slow talking major from Georgia whom Blair had chosen as his second-in-command. 'They're not going to be in any condition for a hike through the Vietnamese jungle.'

'I don't expect them to be,' said Blair. 'Feasibility studies are for discussion, to be adapted. I want us in and out by river . . .' he snapped his fingers '. . . like that.'

'After hitting a camp of that size?' questioned the sergeant of the group, Clayton Lewis.

'All the training has been to kill,' reminded Blair. 'I don't want that forgotten, not for a moment. No prisoners: no one escaping into the boondocks. Just one and the whistle's blown: we've got pursuit and trouble.'

'What about civilian personnel going in and out?' asked Manson.

'There are *no* friendlies,' insisted Blair. 'And that goes for a kid who can just speak and pass on the news to a dozing old man sitting on the stoop who may not have seen us . . .' He paused and said, 'Anyone got any questions on that?'

There was no reaction whatsoever from the men assembled in the room.

'Good,' said the Colonel. 'We'd better go: there's a plane waiting at the Andrews base.'

Despite being allowed supportive involvement for Ninh and Nicole at Camp Peary, which enabled the comforting feeling he knew all that was going on, Hawkins was glad the planning was over. It had imposed upon the couple an even greater strain than the initial, aggressive interrogation so it was fortunate they'd have the opportunity now to rest at Maryland Avenue.

It had, in fact, been a fortunate time all around.

The aerial reconnaissance had been as much a vindication for him as for the refugees and that had come across very clearly from London, as clearly to Harry Jones as it had to him. The timing of the Miami convention was fortunate, too. One of the clearest indications of the newspaper's attitude had been the peremptory way Jones had been ordered to take his place in Florida which meant – apart from the inherently indicated support – that Jones would be for a week at least away from any destructive position in Washington.

Eleanor wouldn't risk another bar. Instead he caught a cab to Columbus Circle and, concealed by people emerging from the station, she picked him up there in her car, driving without any intentional direction towards the river.

'I don't like this platonic relationship,' she said.

'Neither do I.'

'What are we going to do about it?'

'We'll work something out, when I get back.'

'How long do you think you'll be away?'

Hawkins shrugged beside her. 'I've no idea.'

'Be careful,' she said.

'I'm not the one who has to be careful,' he said. 'I shall be safely out at sea.'

'Still be careful.'

'Looking forward to the convention?'

She glanced briefly across the vehicle towards him. 'Don't be ridiculous!' she said. 'You know how I feel about that part of politics!'

'You be careful, then!'

She laughed. 'I expected the drinking warning from John, not from you.'

Eleanor turned the car on Constitution Avenue, running parallel with the Mall and Hawkins said 'I always think of Roman legions parading down these streets.'

'All I ever think of is that every one seems to lead to the White House and that's the only direction in which I'm allowed to go.'

'I love you,' he said.

'I love you, too.'

From Sharon's attitude Patton realised it was a mistake to come to Keans, the off-Broadway restaurant: bizarre even, in the circumstances, because they'd eaten there the night their affair began, but because of that it had become a favourite with them both and he hadn't expected her to be so irritated.

'You've never actually *gone* on a trip before,' she protested. 'Suddenly you're spending days and nights in Washington and now you say you've got to go away.'

'This one's special.' He *had* to go: he'd been prepared not to cooperate if Jordan had opposed him. Thank God the agreement had been so easy.

'Why?'

'It's a government contract.'

'What's so different about this government contract, compared to the thousands of others?'

If only you knew, my darling, he thought. He said, 'I just think I should go.'

'For how long?'

'I'm not sure,' he said. 'I'll get back as soon as I can.'

'You sure there's nothing wrong with the business: after all the money you've been spending on us. On me.'

'Of course there isn't,' he said, dismissively.

'You'd tell me wouldn't you?' she said. 'If there was something wrong, I mean?'

Patton paused, covering the hesitation by sipping from his wine glass. 'Of course I'd tell you if there was something wrong,' he said.

Jordan attended the meeting of the Task Force, prior to going immediately to the Andrews Air Force base for the flight to the Philippines.

'Of course you were right in agreeing to Patton going along,' assured the President. 'It's a bonus he made the request. Now *everything's* parcelled up, so we can control it.'

Chapter Twenty-Two

There were separate arrivals, in an effort to avoid attention, and Hawkins was one of the last to get to Manila, by civil aircraft. The heat engulfed him, as it had in Vietnam, wrapping itself blanket-like around him. He wound the window down and asked for the docks and hoped the driver would be able to make some speed, to create a breeze. He wasn't. The Philippine capital was traffic-clogged and Hawkins sat damp in his seat, staring out unseeingly at the Americanised city.

Rescuing Forest and Bartel and McCloud and the unspeaking Page was the obvious priority – the only priority – but to Hawkins there was another consideration.

How had they been wrong? How had Blair and Patton and Peterson and his father all believed them to be dead, when they hadn't been?

Before leaving Washington Hawkins had read every word of the evidence given at the original enquiry. And then he'd played again the tapes of his meeting with Patton and Peterson and checked the transcripts and read once more his father's account. Always unanimous.

Four passes, Patton had said: as low as he could. *There wasn't a movement. Nothing. They'd been cut to pieces.* Peterson had been equally adamant. *They were all gone. Dead.* His father, too. *There was no life, no movement. They lay huddled together, as if still trying to protect the children who were as lifeless between them.*

Maybe soon he would find an answer, thought Hawkins, as the vehicle finally reached the waterside.

He'd been issued with official documentation before leaving Washington. He was told to dismiss the driver at the dock gates and was taken by a gratefully open and windy American military jeep into a closed-off section of the port through a labyrinth of jetties and sheds to an even more secured section. It had high wire-meshed fences and American marine guards who checked his documentation again and asked for his signature and then compared him against a clipboard photograph which Hawkins was unaware the Americans possessed.

A telephone call was made from the guardhouse and Hawkins was finally waved by, gazing ahead at the grey-painted vessel which formed before him. Hawkins knew nothing about ships but recognised this to be an unusual one. It appeared large, stretching along the full edge of the jetty but the difference was in the superstructure. There was a radar bubble, like Cyclop's Eye and seemingly too big for the vessel, directly at the rear of the bridge area and two separate and again large receiving dishes. Around it all was a Meccano of supports and aerials and as the jeep stopped alongside and Hawkins was able to look into the stern area he got the impression that although it was not marked out as such the space would have been sufficient for the landing of a small-size helicopter. It was named *Elmer C. Gorst.* Hawkins wondered who he had been.

There were further identity checks, at the bottom and then again at the top of the gangway. When he reached the cabin to which he was directed he saw it was a double and that the other occupant had already settled in. Immediately opposite the cabin was a shower cubicle and Hawkins was emerging from it when Jordan came along the companionway, calling out his name.

'Mind me as your double?' said the CIA man.

'Not as long as I get to the shower first,' said Hawkins.

'It'll be cooler when we sail,' promised Jordan. 'Which should be soon. You're the last. That's why I came looking for you: there's people I want you to meet in the wardroom.'

The American waited while Hawkins dressed and guided him back along the corridor and up steps to a higher part of the ship which Hawkins guessed was directly beneath the bridge. The impression was confirmed when they entered a large room which ran the width of the ship, with windows looking out over the stem.

There were three other men already in the cabin but Hawkins stared only at one of them. Until the reason had been explained to him the first day at Camp Peary, Hawkins had been initially surprised at Blair's involvement: he was amazed at encountering Patton. That seemed to be the reaction from the American at seeing him.

Conscious of the reaction between the two men, Jordan said, 'This vessel belongs to the Patton Corporation. Mr Patton does a great deal of work for the government, a lot of it classified. I'm not imposing censorship because I know I can't, not after we've disembarked. All I can do is make a request that in return for all the cooperation you are receiving no mention is made, in anything you write, that will link him to this ship. The supposed ownership is sufficiently far removed officially for there never to be any association and the Agency wants it kept that way.'

'I don't see any difficulty in that,' said Hawkins. He was thinking more of the photograph on the man's Manhattan desk than he was about identifying Patton as a contract industrialist for the CIA; there were dozens of those throughout American business. Did his promise that day still apply? It had been made about a book, not about any newspaper coverage. Here he was primarily a reporter and as a reporter he recognised what he had, with Patton's presence.

'Why wasn't I told!' demanded Patton, who appeared to be wilting in the heat, despite the efficient air conditioning.

Jordan frowned at the demand. 'It didn't enter any conversation,' said the quiet man. 'And you've just got the undertaking.'

'I should have been told,' insisted Patton.

'What difference would it have made?' said Jordan.

'It *is* my ship: I've the right to say who can and who cannot travel in it.'

Jordan's frown deepened. 'A ship fully licensed by the Agency,' he reminded the man. 'Mr Hawkins has shown remarkable cooperation throughout,' he said. 'An agreement of that cooperation was full participation. There's been nothing so far to doubt his integrity.'

Hawkins felt vaguely discomforted at having himself discussed as if he weren't in the room. 'Why don't we talk later?' he invited, knowing Patton's concern and wanting to resolve it, in his own favour. It was the sort of human interest story in which his father excelled.

'You know Blair,' said Jordan, entering the introductions. 'This is Dr Hamilton.'

'Doctor?' said Hawkins, wondering as he queried one title why Jordan had omitted Blair's rank.

'A psychiatric qualification: there's a strong possibility I will be needed if we get these men out,' said Hamilton. He was a bonily-thin man, all joints and edges with an Adam's apple which escalated up and down when he spoke.

Beneath him Hawkins felt the throb of engines as the vessel moved out into Manila Bay on its way to the open sea. 'Yes,' he said. 'I guess there will be.'

'What are the timings?' asked Blair, flat-voiced.

He spoke to Jordan and Hawkins realised that the CIA man was nominally in charge of them within the ship, which Hawkins guessed followed logically from the Agency's sponsorship.

'There's no weather forecast to cause any delay and this is a specially-engined ship: off-coast arrival is scheduled around 22.00.'

'Cloud cover?' persisted the soldier, professionally.

'Intermittent but we're monitoring that closely. We've got direct feed-in to a weather satellite over the Indian Ocean.'

'What about facilities for me on board?' asked the psychiatrist.

Jordan gestured around the room in which they were sitting. 'This is the best we can do,' he said. 'There's a carrier moving in from the Indian Ocean: it's off the Cocos Islands at the moment. It'll be in support position in three days. If everything works and we get them out all you'll be required to do is a preliminary examination to see if we've got any big problems. We can helicopter them from here to the carrier, for whatever further treatment they might need. Then short-haul to Guam, where a hospital plane will be waiting.'

The planning was impressive, thought Hawkins: in the circumstances it had to be, of course. 'What communication facilities are there aboard for me?' he said. 'The news blackout lasts until they're safe. Once they're aboard, the promise is full cooperation.'

'I know the promises,' assured Jordan. 'Once these guys are safe you've got an open wire. And believe me, there's hardly a ship better equipped for communications.'

'I think there should be further discussion about that,' said Patton. 'Between Hawkins and myself.'
'Why?' said Jordan.
'It's a personal matter,' said the other American.

Patton's cabin was a large one and Hawkins presumed he had taken it over from the Captain, as his owner's right. It was on the port side, so that through the reinforced windows Lubang was a smudged bump on the horizon. The water was flat and mirrored in the gold and silver that Hawkins remembered from that other voyage, out to the freighter off Hong Kong. Such a short time, he thought: no more than weeks. It seemed a lifetime; several lifetimes. How long had eight years seemed to Forest and Bartel and McCloud and Page in Vietnamese re-education camps?
'You gave an undertaking!' insisted Patton. He was striding about the room, too upset to sit.
'About a book,' said Hawkins, his argument prepared.
'It still applies.'
'I don't think it does.'
Patton stopped, looking down at him. 'Haven't you any idea what this is like for me?' said the man, pleadingly.
Remembering his thoughts in the taxi from the airport and seeing the opportunity Hawkins said. '*How*? Exactly how did it happen that you were so sure they were dead! I've read what you've said: everything. Always you testify you had no doubt.'
Patton turned away from the question, throat moving. 'Don't you think I wish I knew?' he said, broken-voiced. 'I saw them. Shot up. *Dead*.'
'Just wounded,' corrected Hawkins. 'Just lying wounded.'
'I know that now! Just like I know a lot of other things. Do you know what happened, two months ago! Sharon agreed to marry me. Finally, after eight years of saying she couldn't betray her dead husband, she actually agreed to marry me. We've even bought an apartment in Manhattan: started making wedding lists and fixed a date.'
Hawkins felt pity for the other man. 'Have you told her?'
Patton shook his head, unspeaking.
'Why not?'
'Jordan said it was classified.'

A similar injunction hadn't stopped him telling Eleanor, Hawkins remembered. He said, 'She'll hate you, when she finds out.'

'Isn't she going to hate me anyway?'

Hawkins' sympathy deepened. 'I feel very sorry for you,' he said. 'I really do.'

'Then don't write anything about her and me?' pleaded Patton, turning back to him. 'What's the purpose! Hasn't Bartel been through enough? Hasn't Sharon? What do you think it will do to him when he learns his wife fell in love with the guy that took the helicopter away!'

He'd withheld the details about Nguyen Ninh and been savaged for it by Wilsher and the always-criticising Harry Jones. Hawkins became irritated with himself at the attempted justification. That wasn't an analogy at all. Just over an hour before his integrity was being praised. Where was the integrity now?

'Well?' said Patton urgently.

'I don't know,' said Hawkins.

'Of course you know!'

'Let's see what happens,' said Hawkins. Runaway Ray: a frequent taunt of Jane's. It was strange he should have thought of her, so incongruously. But then it wasn't Jane, was it? Just her mockery. What would Eleanor's advice be?

'Why?' demanded Patton.

'Everything depends upon getting them out, doesn't it?' said Hawkins, succeeding in his own escape. At least this time he hadn't needed a bottle.

Hawkins moved uncertainly through the freighter, trying to orientate himself, wanting to put the question to Blair while the opportunity existed. Patton's answer had had the simplicity of truth about it: an honest, genuine mistake over which the man now agonised, as well he might. It remained, professionally, an enquiry he had to repeat.

Directly beneath the bridge area and the wardroom there was an open companionway encircling the ship. Hawkins emerged on to it, by chance making his way for'ard, not conscious until he was actually overlooking them of the preparations on the wider deck below. The soldiers, in black, undesignated coveralls worked in complete quietness, their rubber-soled boots making no sound against the metal decking.

There were two boats, black rubber and stressed inboard with wooden planking, the rubber extending over the now upswung engines in what Hawkins presumed to be some kind of noise-muffling design. The inside edges were pocketed, obviously in some clearly designated form. As he watched Hawkins saw what appeared to be metal crossbows and capped dart cases placed in matching containers in each craft, then grenades and small rockets in those adjoining. The weapons – seeming stunted or broken down from where he watched – went opposite, in elongated compartments. Spare fuel cans – rubber for silence Hawkins noted, intrigued – were strapped on this side and then more containers already packed with unknown contents, were loaded, each to a chosen area with properly tailored strapping.

Even though it was extremely soft, the first practice ignition of the engines made Hawkins jump. The engines were stopped and started three times and when it was extended over them Hawkins realised he'd been right about the cowling: it cut down the engine sound to an almost inaudible burble.

Hawkins located Blair to the left of the group, watching the work, the Colonel half concealed by some overhanging piece of the superstructure. Hawkins moved on around the companionway, seeking steps, finding them on the starboard side of the vessel which brought him out on the side furthest from the working incursion group. Blair was about fifteen yards away from his men, against the bulkhead and Hawkins called out when he was about five yards away, identifying himself.

Blair turned, coming further out on to the deck and Hawkins began, 'I was wondering, Colonel, if I could . . .' and then trailed away, conscious that the man had been standing with Patton.

'What is it?' demanded the soldier.

'The original mission,' said Hawkins, unhappy at continuing the questioning in front of Patton. 'From what's happened it seems that there was a mistake, in 1975. I was wondering if . . .'

'I told you I couldn't talk about it, in a letter,' cut off Blair.

'I would have thought the circumstances were changed, from the time of the original request,' argued Hawkins.

'They haven't,' insisted Blair. 'And at the moment we don't know definitely that the circumstances have changed: that there are survivors.'

'You wouldn't be making this entry if there weren't good grounds for thinking so,' said Hawkins.

'I don't consider I can discuss that with you either,' said Blair. 'I consider myself bound by the reasons of that first refusal and I don't think anything is going to change to alter it . . .' He looked to where his men were working. 'I further consider that this is a restricted area, irrespective of any promises or undertakings you've been given. I'd like you to leave.'

Hawkins thought it doubtful that Blair had the authority to order him away but recognised the pointlessness of arguing with the man. Determined to refuse him the victory, Hawkins said, 'Perhaps with communication equipment as efficient as this ship has got and with the undertakings I *have* been given I'll get into contact with Fort Bragg again and see if we can change the official refusal.'

Hawkins hurried away before the soldier had time to respond, recognising it as a fairly weak shot but angered at the patronising attitude. He wouldn't do it from the ship, he decided, but he'd try again when they got back to America. He took the ladder again, knowing no other retreat, emerging on to the overlooking companionway and going slightly forward, so that he could look down upon the working soldiers.

'Quite soon now.'

Hawkins started at Jordan's voice close beside him, not having heard the CIA man approach. 'How far are we off?'

'Maybe five miles,' said Jordan. Nodding down to the assembled incursion squad, he added: 'We want to give them the best possible chance.'

Hawkins turned to the American. 'What about the twelve mile limit?'

'It's beyond that we lay off,' he said. 'This thing's got gadgetry you wouldn't believe. There's a metal showering device similar to that used on aircraft, to fog radar.'

'From which any shore installation would guess an intrusion, surely?'

'Surely,' agreed Jordan. 'That's why planes from US bases in Thailand are making over-flights for which we'll formally apologise tomorrow, pleading navigational errors.'

'You haven't missed a trick, have you?' said Hawkins.

'I hope not,' said Jordan. He looked away from the preparation, directly at Hawkins, and said, 'Want to tell me what you and Patton talked about?'

'Like he said, it was personal.'

'I heard what he said.'

'Ask him.'

'I did. He told me to go to hell.'

'I'm not telling you that,' said Hawkins. 'I'm just saying it was personal.'

There was a discernible slowing of the ship and finally it began to lift and fall in the gentle swell, hardly any way on at all. Jordan moved and Hawkins followed, using the CIA man's presence for permission to return from where he'd just been ejected, descending the ladder on to the deck where Blair's men were assembled. Any part of their skin possible to see was cork-blackened, even the palms of their hands, and all wore black woollen berets. Their coveralls, like the compartmenting of the rubber boats, were pocketed and pouched for unknown purposes but Hawkins was surprised to see that they had no side-arms or visible weaponry, only long bladed Bowie knives. He heard Jordan say, 'Don't forget the importance of the timed communications,' and realised there had been at least one briefing from which he'd been excluded.

'Is it likely that I would,' responded Blair sarcastically. He looked at the returned Hawkins but said nothing.

'It's time,' said Jordan. 'Good luck.'

Without replying Blair turned to his waiting men, giving some signal of which Hawkins was unaware. Rehearsed, the men moved the prepared boats easily through the lifted rail, lowering them soundlessly into the water and then using the lowering ropes to make an equally silent but bouncing abseil against the ship's side into them. The engine noise didn't reach Hawkins where he stood, staring down from the rail, watching them move away.

'I'd like to be as brave as those guys,' said Jordan, from beside him.

'Why weren't they properly designated as soldiers?' asked Hawkins. 'Those suits were entirely unmarked.'

'It's part of their bravery,' said Jordan.

Chapter Twenty-Three

Timing was crucial. There are six main exits comprising the Mekong Delta, two of them plugged by middle-river islands. The tributary leading to Can Tho is one of them. There are, in fact, two separate strips of land blocking the way into the South China Sea. The rehearsed and exercised plan was for Blair to take his group below Cau Tranh De, the southernmost obstruction, through a channel a few hundred yards wide but before 3 am, the time the Americans knew from their earlier presence in the country that Vietnamese fishermen sailed for their day's catch.

They travelled roped together at a distance of about ten yards, without any illumination, Blair in the lead boat giving signals to his own crew by hand and muttering them softly – little more than a voice vibration – into a throat microphone to be relayed to the headset of Manson, in the following craft. Despite the importance of time, they initially moved at the slowest speed necessary to counteract the tides and still make headway, to reduce the cowling-lowered engine noise further. It made accurate navigation difficult, with the need for constant compass checks to ensure they were still on course. Manson had a matching setting so he didn't have to rely entirely upon the instructions from Blair. Apart from the Colonel no one spoke in either boat; they knelt against the rotund gunwales, staring in the direction of the land they could not see, every man a look-out.

The actual incursion had been one of the major exercises on the York River, at Camp Peary, the proposed journey precisely measured from the gauged point of their drop-off from the *Elmer C. Gorst* to the river mouth and equally precisely timed.

What had to remain an estimate was the tidal resistance and within an hour Blair realised there had been a miscalculation; what should have been a black, humped horizon of approaching land remained flat, sea-edge against night-sky. A necessary delay factor had been built into their timing but Blair was reluctant to encroach into it so soon. He muttered the order to the following boat and increased his own speed, bubbling a white wake behind him; if they were that far away they could afford noise. Certainly afford it more than confronting a Delta fishing fleet, on its way out to sea.

It was a further hour before the skyline broke, initially identified by the man at the very prow and confirmed immediately through infra-red nightglasses. From the land sighting Blair knew he had eroded thirty minutes of his failsafe time. The communication between himself and the man with binoculars was entirely signalled: the man looked back, through the glasses, now seeking the break that would be the river mouth. The signals came back that the landmass was solid and Blair confronted another miscalculation, this time navigational. He changed course, northwards, still risking the faster speed.

He was too professional for any alarm; so were the men about him, aware from the intense training what had happened. The first indication of the tributary was physical, rather than visual, an increased rocking of the boats as the outcoming currents, speeded by the bottleneck of the islands, fed into the sea. The look-out's indication came as Blair throttled back. To combat the current it was still necessary to keep on more speed than he had intended: the cloaked engines gurgled behind him, sounding loud.

Blair abandoned his compass, relying entirely on the left-right-left gestures from the man with the night-seeing binoculars, twitching the tiller according to the instructions to keep the tiny convoy absolutely in mid-stream. The run was faster, directly between the island and the bank, not just resistance at the head but jostling side-currents, which meant a constant jerk and then correction upon the steering as Blair attempted to keep on course. He had a positive position, between island and shore, and from it recognised he was ninety minutes into his emergency time; too much to maintain the original schedule and try to get beyond Phung Hiep.

It was an irritant, nothing more. He was sweating with the effort of keeping the boat on course but uncaring, looking intently into the darkness for the first sight of the jungle and foliage he knew so well. Already he could smell the familiar smell, like a man coming home after an absence to a familiar mistress.

Blair was conscious of the race decreasing and realised they had cleared the island to their right, entering a less restricted and therefore more sluggish part of the river. Gratefully he throttled further back and relayed to the following boat the changes necessary because of the delay: he didn't bother to turn, to get confirmation from Manson.

Blair knew there would be hamlets along the bank established since the time of the American maps he was carrying and accepted their risk but Long Phu was a sizeable town by Vietnamese standards, on the left in the direction in which they were travelling. He steered against the far bank, to skirt it as widely as possible, wanting to minimise the danger. Blair decided objectively that they had done well: damned well. They'd lost time but they were still well within their limits, so it was not a difficulty; it was basic military lore not to expect an actual operation to retain the schedule of an easier exercise. All right, so they would have to stop short of the original objective, concealment beyond Phung Hiep, but it was only by a few miles, which was unimportant. Still damned good.

From the front came the warning of the township on the far bank and Blair whispered it to the following Manson and took the engine down to its lowest note, only just moving the boat forward through the water and edging it, too, closer to their side, close enough easily to hear the birds screeching surprise at the coming day and the hidden jungle sounds of pushed-aside undergrowth. Despite the darkness they could see the jetties directly opposite pointing accusingly at them and beyond the moorings the regular shaped outline of houses: against the just lightening nightsky an unidentifiable tower, maybe a clock or church, stood erect, another reproving finger. There was an occasional bark from a dog showing its bravery but otherwise the Long Phu slept undisturbed.

Two am, Blair saw. Time to conceal themselves beneath the overhang of a bank, out of the way of any fishing boat on its way downriver from Phung Hiep and out of sight, too, of any awakening hamlet. It was going to be a boring day. He passed the order back to Manson and then hand-spoke the instruction to the man with nightview: the glasses swung from bank to bank, seeking the protection. From either bank more birds were becoming excited and the sky was lifting to a positive grey, so they could see the tree line against it. Two-twenty: if there were not something reasonable soon they would have to take what was available and improvise upon it.

The gesture came, towards the far bank and Blair looked towards it through his own binoculars, nodding approval at the choice. On either bank the lush vegetation crowded down actually to the water's edge and often overhung beyond but the look-out had located a better-than-average canopy, a great bulge of covering branches and shadowing leaf beneath which would be a completely hidden tree cavern. The instruction to Manson and to the soldier in his own boat was simultaneous: Manson closed up and the soldier took in the umbilical rope, shortening the distance between them. The river push held them apart, so that they approached the concealment not one behind the other but side by side and it was the odd formation which compounded the accident.

There was no break in the water, to warn of what must have been a splintered, current-sharpened tree branch, mud-embedded just below the surface. There was a jolt, of immediate contact with Manson's boat, and then a squeaking, tearing sound as the momentum of the craft carried it forward, slicing open the air-filled port side from stem to stern. The rubber was reinforced and designed against complete collapse but the opening was so large and so abrupt that the deflation was dramatic. It threw Manson's boat violently over to the left and because the tethering was so short between them that action caused Blair's boat to tip in the same direction as well. Four men went overboard immediately from Manson's boat, two from Blair's and within minutes Manson's boat settled further in the water, submerging the occupants and dragging Blair's craft so far over with it that water began shipping over the side. It was Blair who reacted, snatching out the Bowie knife and severing the connecting rope between them. Relieved of the strain, his own boat righted itself but without support Manson's craft flopped deeper.

Risking the sound, Blair shouted 'Don't let it drift away: bring it to shore.'

Men arranged themselves in a pattern around the damaged boat and on Blair's orders Sergeant Lewis, who was travelling with him, threw them a longer line. Risking sound again because it was the only way, Blair increased the power and strained forward, towing the waterlogged boat and its clinging occupants on towards the tree protection. In Blair's boat men felt out, groping for the approaching trees to drag themselves into their cover and when they found it was sufficiently shallow four went over the side and hauled manually on the rope, bringing the other soldiers into shore. Beneath the trees it was utterly black again and as they waded ashore there was the scurry and slither of displaced occupants.

'Roll call,' ordered Blair, at once.

Obediently the squad recited their names. There was the briefest pause and Blair demanded, 'Jackson? Where's Jackson?'

There was no reply.

'Andrews, Jones, secure the boat. Everyone else back to the tree edge,' ordered Blair. There was no panic in his voice: emotion even.

The seven men re-entered the river, wading and swimming to the point where the protective tree skirt met the water, snatching up for the overhead branches when they got there to support themselves against the current. It was approaching near-light now, the sky's greyness already giving way to the orange and pink of an exploring sun.

'There!' said Manson.

About twenty yards downstream Jackson's body was caught against a tree arm, some part of his webbing or tunic snagged on an unseen spur. The force of the water had spread his arms against the trunk, as if he were sacrificed and the weight of his body jostled by the force of the water, caused it to rise and fall, submerging his face, then raising it, then submerging it again. He was quite exposed in the open river.

'Get him back, quickly,' ordered Blair. To the sergeant he said, 'You, Lewis. Secure line.'

They tied the rope around the sergeant's waist and then fed it through one of the larger and stronger forks above their hands, to provide a fulcrum. Lewis merely let go of the tree and let the current carry him towards the dead soldier, both hands raised to grab the trunk against which the man was impaled. He caught it first time and beneath the protecting tree the other Green Berets immediately took the strain, holding him by the rope against the downthrust of the river. Lewis began edging sideways along the tree arm and then stopped, appearing to hear it at the same time as the rest.

There were three boats in the tiny fishing fleet, being moved easily with the downstream run by the back and forth movement of the steering oar, the bird-song Vietnamese of the crew shrill in the still morning.

'Set out like targets on a funfair shooting gallery,' said Manson, of the sergeant and the dead Jackson.

'Ready to take them,' ordered Blair, aware it would be the end of the mission if he had to. Four men waded back to shore and returned at once with weapons, training them upon the boats gradually rounding the bend in the river. 'Wait . . .' said Blair, warningly, seeing Lewis move again after the momentary hesitation. The sergeant continued hand over hand, groping out for the dead man to detach him from the spur. Blair realised what the man intended, and just had time to warn the others besides himself still holding the rope, before Lewis dropped into the water, arms and legs wrapped around the corpse, letting its weight drag him unseen beneath the surface. The immediate pull upon the rope and the men holding it – Blair, Manson and a soldier named Penley – was enormous, so great that the fulcrum branch dipped low into the water, threatening attraction by its very movement.

'Two back here,' gasped Blair and half the ambush squad grabbed for the rope, lessening the tension.

Out in the river the Vietnamese were in full view now, seine nets folded, no one working except the steering oarsman. They seemed disinterested in the shuddering tree to their right; Blair guessed about ten feet of the taut supporting rope would be visible, if they looked.

'Pull him in,' he said. 'Quickly.'

They began to tow in unison against the rope, dragging Lewis and literally the dead weight of Jackson against the river flow, necks corded with the effort. About ten metres from them Lewis, unable to stay submerged any longer, surfaced briefly for air, still clutching the body: the fishing boats were about twenty metres beyond him, downstream. No one was looking back.

When Lewis got to the protection of the trees others took Jackson's body while Penley and Manson supported the choking, gasping sergeant. As they got him near to the bank, Lewis was sick.

There was no acknowledgement from Blair for what the man had done.

Everyone was soaking wet and mud-smeared and the slime increased when they got to the bank, too shaded for any grass to grow. No one tried to avoid it or clean themselves up because it was useful. It stank, the smell smothering any human odour which might attract any camp or village animal and raise an alarm.

'What's the damage report?' demanded Blair.

'The boat's useless, in these conditions,' said Manson. 'There's still some reserve buoyancy in the port float and theoretically it could still be used but in this current it would drag: couldn't be manoeuvred or driven without the engine at almost full power. Whatever it was cut through the inner lining of the container pouches too: took the rubber protection off the reserve radio so that's probably water-damaged.'

Blair kicked the largest container still strapped in his own craft and said, 'So there's this reserve boat which we wanted for the guys in the camp. And mine.'

'That still gives us enough room,' said Manson. 'There's nothing wrong with the engine of this boat: it's just the skin.'

'Enough room providing we get there intact and don't run into some other kind of obstruction,' qualified Blair. 'It's too risky.'

'What then?' asked the major.

'We can't lay up and use the river,' decided Blair. 'We'll stash the two remaining good boats here, high up in these trees, backpack what we can carry and move on right away, through the jungle. Phung Hiep is the only town of any size we've got to skirt.'

'This is rice paddy country,' said Manson. 'There'll be dozens of hamlets we don't know about.'

'I know the country,' reminded Blair, pointedly.

'We've got a body,' reminded Lewis, recovered now.

'Take the tags, everything personal. Bury him here, in a body bag. I want to move in an hour.'

There was no reaction from anyone at the callousness: Jackson was dead and they were alive. Their Commanding Officer had taken the best option to remain that way. That was how they were trained to exist. *To* exist.

While two men stripped the dead Green Beret, body-bagged him and then dug the burial spot in the soft mud, the others unloaded the boats and secured them both high into the branches of the tree, far above the highest flash flood, and camouflaged them and their binding with foliage. By the time they finished the grave had been covered and washed over with scooped river water, to smooth away any trace of digging. Blair paused briefly at the unmarked spot and said, cursory but bowed-headed 'May he rest in peace' and then carried on into the jungle.

It was tangled, matted undergrowth, their feet sinking ankle deep in unseen water and slime and they moved slowly because there had to be no trace of their passage. There was heat in the day now and mosquitoes and flying things swarmed around, eating at them and stinging. It was a long way, almost a mile, before the dense-packed tangle cleared slightly. Immediately Blair despatched Penley, a New Yorker with a film parody of a Brooklyn accent, and a negro named Wright, from Alabama, to act point, ahead of them. To allow them to scout, the others rested.

'How far do you think?' asked Manson. He caught a long-legged insect on a eucalyptus leaf with a squirt of tobacco juice.

'By going straight we're taking the curve out of the river,' said Blair. 'Maybe thirty miles, in a direct line.'

'Which we won't be able to maintain.'

As if on cue, Wright returned to say that within five hundred yards the paddies began, wide, open areas of rice cultivation through which they couldn't possibly pass without being seen.

'River line seems best,' said the negro.

'Back to the curve,' said Manson. He tried another insect but missed this time.

'There'll be cover,' said Blair, showing his knowledge of the country. 'It'll be risky but there will be cover.'

He led out, after Wright, coming to the place from which

they could see the paddies, flat and glittering in the morning sun to their left. Cone-hatted peasants, like human mushrooms, were already working, bent double. Near where the Green Berets crouched the ground was banked, a dyke between the cultivation areas and the actual river which irrigated it and it was towards this man-made obstruction which Blair gestured. To keep the sluices clear, undergrowth was cropped back on the river side for a distance of perhaps three metres, providing a comparatively clear path completely concealed by the banking from the workers on the other. And foot trodden, so there would be no evidence of their passing. Penley and Wright still formed point, more directly now, moving out to the first sluice gate to ensure it was clear and then signalling the remainder on: they could run quite easily and unobstructed between the stopping points. Heat filled the day now, pressing down on them: the only sign of effort from the superbly trained group were the black patches of perspiration beneath their back-packs and their arms. At midday Blair took them further towards the river, into the intervening jungle and announced a fifteen minute break.

'This is going good,' said Manson.

'I'd like to make Can Tho by nightfall,' said Blair. 'Sooner if possible.'

'*We* can travel like this,' said the major, warningly. 'If there are guys who've been imprisoned for eight years, they won't be able to.'

'I know,' said Blair.

'Haven't we got a problem?'

'The camp's river served,' reminded Blair. 'We'll use their boats.'

'What if there aren't any?' persisted Manson.

'We'll wait until there are.'

They got to the outskirts of Phung Hiep an hour later. The approach was marked by the end of the paddies, a separation between the township and the low lying water-area of low scrub, with an occasional tangle of deeper jungle through the middle of which ran the high raised, American legacy road.

'Known better,' said Manson, crouched alongside Blair.

'And worse,' said the Colonel.

They watched for a long time, gauging the use of the road and assessing the cover available. There was a spine of heavier undergrowth about fifty metres from them, running up into the coppice bordering the road but the space separating them was completely open, a paddy split only by a single-track dividing one work area from another. Manson went first, belly down in the water, hauling himself through the mud and using what narrow protection there was from the walkway. The rest took cover positions but, as he had when they prepared to fire on the fishing boats, Blair recognised that to shoot just once, until they got to the camp, would end the incursion. Manson reached the protection, a spiny upthrust of bamboo, and reversed the guard stance, for the rest to follow, one by one. Blair was last. They edged deeper into the tree and brush tangle, in the direction of the road. It was raised, against flooding, stormditched on either side, so they did not have a completely clear view in either direction and certainly couldn't see anything in the opposing ditch. Penley acted as scout again, stripping himself of everything and scurrying out to lay flat against the ditch edge, able to look in both directions along the road. At his signal, Manson moved first again, running doubled to where Penley lay, pausing momentarily, then hauling himself over the lip of the road to cross to the other side.

The Vietnamese was resting. His bicycle was neatly beside him and he squatted, in the Asian fashion, comfortably upon his heels. The whimper was of surprise because there wasn't time for fear as Manson hurled himself into the ditch. And then his mouth opened wider but he never had time for the shout, either, because the American chopped upwards with a motion that carried on from his landing, catching the peasant's chin perfectly with the heel of his hand, snapping the man's head back to break his neck. Without pause Manson turned back to the road, positioning himself as Penley was on the other side, giving them complete visibility of the road. Again the crossing was singly: from the ditch every soldier carried on directly into the bordering jungle, ignoring the neck-twisted Vietnamese. Only Blair, the last to cross, stopped. And then only briefly. He looked at the Vietnamese and then at the bicycle. He pulled the machine nearer and brought his heel down against the front wheel, as if it had buckled in collision and then threw it across the body.

To Manson he said, 'That'll be good enough.'

Back in snagging, clothes-pulling undergrowth and jungle

their progress was slower than it had been along the dykes but Blair forced them on, careless at times of leaving traces, worrying only about ridged bootmarks in the soft ground and dismissing broken bushes and branches as explainable by animal damage. They skirted Phung Hiep in a wide arc and returned towards the river, although this time they didn't attempt to reach the dyke path but kept the paddies between them and the waterway. Twice their way was barred by hamlets, which involved wide skirting movements, but by late afternoon, when Blair halted them, they were within two miles of Can Tho.

'Well done,' he said, his first praise. 'We'll make it tonight.'

The satellite co-ordinates put the camp short of Can Tho, about a mile downstream from the town. Once more the paddies ended, as they had at Phung Hiep, and this time they recrossed the linking road without encountering any Vietnamese. Scouting was vital now, not just for their protection but to locate the camp and Blair sent out two more men, another negro named Abrahams and a squat, rarely-talking man called Lewin, to aid Wright and Penley. It was Lewin who found it, within five hundred yards of the satellite-fixed position. They regrouped and in tight formation moved around to the best jungle cover, to the west. To their left, north east, the widened tributary arced yellow and brown past the expected jetties, alongside which were moored two small craft and a larger, flat-bottomed supply boat. Beside the immediate warehouses were smaller sheds, a thin, straight roadway and then the camp constructed exactly as it had been described by Nicole Tiné and Nguyen van Ninh. Beyond the camp, to the south and east, were the workfields, paddies and then better drained areas, for vegetables. Outside the camp, to the east, there was a straggle of thatched huts for outside workers employed there.

Normally this part of a reconnaissance would have been left to the Commanding Officer but they were looking for imprisoned Americans and no one relaxed, scouring the area through field-glasses protectively shielded against tell-tale glare from the late sun which might have disclosed their position.

'By the far dyke, near the oxcart,' said Lewis.

It was a straggled line, three shuffling, automatic men carelessly guarded by two Vietnamese who gossiped behind, smoking and laughing and almost completely disregarding their charges. Unspeaking, the American assault team watched as they filed dutifully into the camp, past the outer and middle perimeter and into the core of the settlement, disappearing into a hut to the right of a large building.

'Our guys,' said Manson, empty-voiced, as if he were unable to believe it.

'Forest, Bartel and McCloud,' identified Blair.

It was Patton who brought the news from the radio room and for the first time Jordan's control went. He leapt up, seeking expression, finally seizing the psychiatrist's hand and shaking it, as if the man deserved some sort of congratulation.

'Son of a bitch!' said the CIA man. 'Son of a bitch!'

Hawkins was more subdued, recognising the attitude as selfish but allowing it anyway, his mind occupied by what it meant to him. It still meant unresolved doubts – perhaps more – but immediately it meant he was completely vindicated, in everything he had done. And that he was closely, intimately, involved in a world-beating exclusive as dramatic as anything his father had known. Perhaps most important of all, he didn't feel the slightest need for a drink, to celebrate.

Hamilton, the psychiatrist, also curbed any excitement. 'We've positively found them,' he said, cautiously. 'We've still got to get them out.'

Abruptly, without saying anything, Patton turned and walked from the wardroom.

Much later, in the darkness of their shared cabin where Hawkins thought the CIA man to be asleep, Jordan suddenly demanded, 'Why did Patton leave like that tonight?'

'I don't know,' lied Hawkins.

'Anything to do with private conversations?'

'Goodnight,' said Hawkins. He found it easy to sympathise with Patton. The last few hours had been some of the most intensely filled and exciting he had ever known but despite the dramatically radioed news and his complete and sober awareness of what it meant to him, professionally, there had always been a part of Hawkins' mind occupied by thoughts of Eleanor, wondering what she was doing and who she was with and if she were thinking of him as he was thinking of her. Christ, he loved her so much.

The convention, like American political conventions always are, was a circus for the enjoyment of the performers, the acting-out of decisions already long ago decided behind anonymous committee doors. The predictable speeches were greeted with the predictable enthusiasm, the bannered delegations made declarations and achieved their fleeting moment of nationally televised glory, the balloons cascaded from their restraining nets and the souvenir sellers made a lot of money.

John Peterson's official adoption as his party's Presidential candidate was unanimous and ecstatic. There was the proper, necessary reception and the obligatory suite parties at the Fontainebleau and the crowds were still chanting outside in Collins Avenue when he and Eleanor finally reached their apartments.

'I've got it!' said Peterson. 'I've got the campaign and I've got the financial backing and I've got ratings way ahead of Harriman. This close, I'm not going to lose it!'

'Then be careful,' she warned. 'I don't think you're close at all.'

Chapter Twenty-Four

The reconnaissance, like everything else, had been practised before they left Camp Peary, so Blair's instructions were brief. A group under Sergeant Lewis reconnoitred the riverbank and jetty installations and another squad went inland, surveying the prisoner-worked fields and paddies, trying to estimate the guard strength. Blair and Manson remained at their original observation spot, watching and assessing. Just before nightfall the main body of prisoners returned to the camp, the difference between the Vietnamese political and criminal detainees obviously marked. The guarding of the political section was almost as casual as it had earlier been with the Americans but with the true criminals it was tighter, armed men spaced out at ten to fifteen metre intervals. As the men filed into their prescribed divisions inside the main wire, general illumination right around the stockade came on and in it the watching Americans could see the guard change at three guard towers. Searchlights flared on individually from each, testing, and then died again. After about an hour there was a recognisable shrill upon a siren but the men lounging in the outer perimeter area made no response. Beyond them, Blair and his men saw the Vietnamese political prisoners dutifully filing deeper inside the camp, into the main lecture hall. There was still no sign of the Americans who had gone inside the bordering hut, hours before.

The squad who had watched the fields returned first: Blair heard them – just – but only in the very last moments and nodded approvingly as they entered the enclave he and Manson had created.

'We counted twenty in the fields,' said Penley, spokesman

for the group. 'Maybe ten more came out to increase the escort back.'

Blair nodded and said, 'Our calculation, too. Let's estimate another twenty for roster changes and split duties. A total of fifty then.'

'It's enough but I don't think the assigned guards are the main problem,' said Manson. 'With surprise we could take them out. It's this outer cordon of prisoners I don't like. There have got to be five hundred, maybe more, that we've got to get through.'

'That's the difficulty,' agreed Blair.

'I don't see any particular electronic protection, in the camp,' said Andrews, surveying it with nightglasses.

'They're complacent; soft. That's in our favour,' said Blair.

There was the faintest rustle and Lewis led his group in, hunkering down close to the officers.

'There's a guard post near the jetty but we watched for an hour and nobody bothered to man it,' he reported. 'One of the smaller boats is motorised and could take everyone: no idea about keys, of course, but we could jump the ignition easily enough.'

Inside the camp there was another shrill on the siren and this time the outer assembly reacted, moving towards the mess-halls: nearer the centre, the Vietnamese emerged from the indoctrination area, making towards the same eating area.

'There are our guys,' reported Andrews, still with the binoculars to his face.

Everyone looked. There was a fourth man now, black, walking with his hand against the shoulder of the American in front, McCloud. They shuffled towards the dining barracks, prodded by only one guard.

'Motherfuckers!' said Wright, another black, quietly. 'I want me some motherfuckers.'

'It's not personal,' warned Manson, with equal quietness. 'No one ever fights well when it's personal. Stay cool.'

'Let's get it together,' said Blair, bringing the men back to him. 'They're buffered by that outside cordon of prisoners. No frontal assault then. We're going to take them tonight, complete surprise, in and out of that last perimeter without anyone knowing we've paid a visit. We'll take the motorised boat and immobilise the other two. Andrews?'

'Sir?'

'Communications. I want everything taken out, every wire cut, every radio and transmitter destroyed. But quietly. No explosives. Rig it though. Leave one radio apparently undamaged and booby-trapped. When they start signalling I want them blown away. Abrahams?'

'Sir?'

'Rig the camp behind us. Main guard room, barracks, everything. Whatever they touch, whatever they move, wherever they run, I want it to go up in their faces. Anything and everything to delay any pursuit. Wire the jetties too. I want them gone, as well as the other two boats . . .'

'They've got a system, the sneaky little bastards.' The interruption came from the already-briefed Andrews, who'd gone back to watching the camp through his nightglasses. 'Look,' he urged. 'Torch flash, from tower to tower. I've counted it out: average every six minutes.'

'Well done,' congratulated Blair. 'Wright, Penley and Yates. When the briefing's over you'll synchronise a time. Let the light go from the tower nearest us here: that gives you six minutes to scale all three and take them out . . . no sound.' Blair hesitated and said 'Which leaves Forest, Bartel, McCloud and Page . . . I'll do that: hopefully Forest will recognise me. React properly. In case he or the others don't I'm taking in straitjackets and shots, to knock them out . . .'

'Then you won't be able to handle four men by yourself,' pointed out Manson.

'Initially by myself,' qualified Blair. 'I want my back covered, every step of the way. But I don't want a rushed entry into their barracks. They're going to be disorientated enough, without imagining some sort of attack . . .' He stopped again, thinking. Then he said, 'Remember, one shot and it's all over. I don't care what happens once we're aboard their boat but I want us out of that camp before the noise starts.'

'We can't guarantee that,' said Manson, objectively.

Blair nodded, accepting the argument. 'Penley?'.

'Sir?'

'You take the tower nearest to us here. Stay there. See us away. If there's any shooting and the outer perimeter people come out, you've got the gun and the elevation. Take grenades, too: grenades will be more effective, at first. Any thoughts?'

'What about delayed fuses on the boats and jetty?' suggested Manson. 'If they go up in mid-river, following us, it might cause a blockage and hold up any pursuit from the town of Can Tho itself.'

'Good idea,' accepted Blair. 'Got that Abrahams?'

'Yes, sir.'

'We'll take the towers at midnight,' decided Blair. 'Rest.'

And they did. Such was the degree of their training that despite what they were going to be doing in a few hours every man was able to relax, not beyond the awareness of what they might confront but able to compartment it. Half an hour before the appointed time Wright and Yates stirred, with ground to cover to get into position on the far towers. They synchronised with Penley and eased away into the darkness. After fifteen minutes Blair ordered 'Now you,' and Penley moved off, to the left, his attack already decided.

The camp lights were concentrated inwards, towards the barracks and the exercise areas, the space immediately outside the wire neglected. What brightness there was was cut into a wedge by the shadow of the tower itself and it was this the Green Beret intended using. Through their nightglasses they watched Penley ease down into the ditch that ran alongside the roadway from the jetties and scurry along it until he reached the black triangular shadow. He entered it, momentarily pausing for challenge and when it didn't come dashed to the base of the tower, thrusting beneath the struts. There was a brief time check and then he craned up, waiting for the torched signal. As soon as it came, he leapt for the criss-cross supports: back among the main body Abrahams was not looking at the ascent but directly down at his watch, timing off the climb.

'Five . . . five and a half . . . four . . . four and a half . . . three . . .'

'He's there,' stopped Manson.

There had been a pause, just below the lip of the tower platform, and then Penley had gone over in one fluid motion. There was no sound. The group waited, looking beyond their immediate tower in the direction of the other two. Almost at once, from the one furthest east came the signal, not one but two torch flashes, matched at once by that next in line and completed by Penley.

'Let's go,' said Blair, quietly.

They ran, doubled-up, one at a time across the intervening ground to the main gate, grouping around the side supports while Manson checked for any electronic alarm. There were three, cleverly placed, one activated to sound if the gates were thrust open and two more linking a thread of wires on either side, making a snipped entry or exit impossible. They chose the one to the right, clamping bypass leads with alligator claws around the system and then cutting their way in, not through a narrow hole but – conscious of having to lead confused people through when they left – with an easily passable gap. The criminal barracks were elevated, on stilts, and they used the gap for concealment, conscious of the shift and stir of men immediately above their heads. They crabbed beneath the full width of the barracks, and still hidden were able to look out towards the next encirclement.

'Shit!' said Manson, softly.

There was a guard post by the gate into this second section and obvious occupation although it was impossible from where they watched to guess how many people were inside. As they watched one emerged, unzipped his fly, relieved himself against the edge of the box and returned inside.

'What about guard changes?'

'The woman didn't know, not accurately,' reminded Blair. 'We'll take them out and leave one behind, just in case. Lewis.'

The sergeant selected another man by tapping his shoulder and together they eased away from the barracks, keeping still to its shadows to a point where it almost abutted the wire and then returning parallel, on the blind side of the guard box. The flash of the overhead light against Lewis' knife could be clearly seen from where the others hid as he drew it but it was a bad attack by their standards, the startled cry from a surprised sentry faint but loud enough to reach them. They tensed, heads tilted sideways for the first indication of a reaction from the sleeping men above their heads: two of the Berets behind Blair had their hands on grenades, in readiness. There were vague sounds of movement but nothing more.

Lewis had already found another set of electronic alarms by the time they reached him. As Blair stooped alongside, waiting for Manson to immobilise them as he had on the main gate, the sergeant said, 'Sorry. One was asleep just inside the door: I didn't see him.'

They used the barracks for concealment again, this time that of the Vietnamese political prisoners, and from its protection they were able to see clearly the main living quarters of the camp guards.

'It isn't going to work,' said Manson, always objective.

The building was directly in their path to where the Americans were housed. Some of its windows were still lighted and there was movement and cigarette glow from the wide verandah which surrounded it; distantly came the reedy music of some wailed pop song.

'We've got to take them,' agreed Blair. 'Crossbows, garrotte and knife: no matter how good Penley is in that tower, we'd never be able to get back through all those behind us.'

Lewis, conscious of the implied criticism from the guard-post attack, led out, metal crossbow – an adaptation of the medieval weapon most commonly used by modern-day commandos – cupped across his body, bolt already fitted. Three others followed. Blair and Manson remained where they were: Manson had a short-stocked adapted M-16 unslung and ready, Blair a silenced Browning automatic. The Vietnamese on the verandah were grouped together, four of them. This time the assault was perfectly coordinated: Yates and Abrahams came in from behind, taking the two on either extreme with knives and Lewis and Wright jerked abruptly upright in front of them, killing the inner two instantly with crossbow bolts. There was no cry, but although Yates and Abrahams tried to cushion their collapse, there was still the sound of falling bodies: the four Americans spread themselves either side of the door, tensed for any investigation. One reedy tune finished and another began: otherwise nothing. At the gesture Manson and Blair crossed, the others following.

The entry did not lead directly into the barracks. Immediately inside were doors to what Blair guessed were offices and on the other side a washroom and toilet area. He nodded to Lewis, who eased in pressed against the wall, going along the short linking corridor to look into the dormitory section. He was back within minutes.

'Eighteen that I can count,' said the sergeant. 'Guy with the radio is three along on the left, just lying in his cot. There's a game of some sort going on ten cots down to the right: three guys. Everyone else appears asleep but I can't be sure.'

'Yates,' said the Colonel. 'Take the man with the radio. Lewin, Abrahams, Wright, those on the right: one each. Those asleep as they lie, all of us.'

The deputed men went in first, the sergeant's group leading with Yates directly behind. The initial assassination went without sound but two of the game-playing group fell against adjoining beds: before the occupants realised what had happened Abrahams and Wright were upon them but there was a sound which roused others. Three at the far end of the hut scrambled up, bulging-eyed and momentarily dumb with terror and Blair killed them one by one with the silenced pistol, the only sound its soft 'phut'. One Vietnamese not asleep tried to grapple with Wright, crashing over with him, and the Green Beret let the man actually fall on to his knife. There was never a pause in their action. The moment the main hut was cleared, Lewis and Yates ran back to the entrance, checking the closed doors in the corridor. They were offices but they were empty.

The group assembled in the corridor and Blair said, 'Only half, according to our estimate. Abrahams, start wiring. Everyone else, watch my back.'

Blair stopped at the door, checking outside and then dashed in the direction of the building the Americans occupied. He was about ten yards away when from an unseen guard post completely concealed in the shadows a Vietnamese appeared, raising a rifle as he came and beginning a shout of challenge. Blair didn't pause, actually running faster and kicking out as he neared the man, a brawling, brutal, bone-cracking kick into the groin. The man doubled, screeching the pain as the breath was driven out of his body, dying immediately as Blair's knife came down into his back. Blair flattened himself against the wall of the hut, back tight against it, staring out into the camp. Two Vietnamese came running from a small building set apart from their main barracks, looking around startled and never seeing the cause of the noise because crossbow bolts fired by Wright and Lewin on the verandah killed them instantly.

Blair didn't wait.

Confident of their outer security, the Vietnamese only bothered with a simple lock on the door which gave within seconds to pressure from Blair's thick-bladed knife. Inside the room was in complete darkness and utterly silent. He ducked in low, presenting no profile, moving immediately sideways.

Still nothing moved. With the knife point he prodded the door, closing it, raising himself from the crouched position but not moving, to disclose his position. Just deep, even breathing. Knowing, because of the darkness, that the windows were covered Blair felt out light fingered for the switch. He located it, hesitated and then plunged the room into sudden brightness, the silenced pistol swung two-handed before him in readiness.

There were four cots, two facing two. In the cots, regimented upon their backs, lay the four Americans, Forest and McCloud one side, Bartel and Page the other. All were covered, to just below their chins, none with their arms beyond the covering; but for the breathing, Blair could have been in a small mortuary. He went first to Forest, gazing down. The face was thin, fittingly cadaverous, deeply etched with lines.

'Colonel Forest!' The obedience to once-superior rank was automatic, without thought. 'Colonel Forest!'

There was no reaction from the comatose man. Hesitatingly Blair felt out, for the man's shoulder, ridged and bony beneath his touch. 'Colonel Forest!'

Silently Blair looked beyond, to Bartel – conscious of the baldness and the scar that Nicole had spoken of to the enquiry – then back to McCloud, whose eyes twitched, in constant wincing expressions, even asleep.

He sighed, deeply, snatching into his pocket for the prepared hypodermics. He had the cap off the first, Forest's arm already exposed along the bed covering when the voice behind him said 'Why!'

Blair turned, to see Manson.

'They're shot now,' said the major. 'Why dope them any more: they've got to walk. To the boat at least.'

'It'll be a shock, when we bring them round.'

'Let's bring them around first,' urged the man.

Blair hesitated, then recapped the syringe. 'Let's try,' he said.

Manson got on the other side of the unmoving Colonel, shaking him more roughly than Blair had attempted. It took a long time and then Forest's eyes opened. But that was all that happened, just an opening of the eyes, with no attempt to look either left or right or even move his head, to discover the disturbance.

'Sir!' said Manson. 'Colonel Forest.'

The man moved, but only to the noise and looked expressionlessly at the major.

'He doesn't know who I am: recognise me as an American!' said Manson, disbelievingly.

'Colonel Forest!' said Blair. When the man didn't move, Blair felt out, physically moving his head, so that the man would look at him. The eyes were absolutely empty and blank. Suddenly imagining blindness Blair waved his hand urgently before the man's face and he blinked.

Manson looked at Blair across the bed and said, 'What the hell's happened? He's a dummy.'

'The others,' ordered Blair. He carried on to awaken McCloud and the major went to the opposite side of the room, shaking awake first Bartel, then McCloud. Blair heard Manson say 'Jesus Christ,' and said 'McCloud too: they're all the same.'

'Let's get the poor bastards out of here,' said Blair.

Manson summoned Lewis from his watching position directly outside the barrack and with the awkwardness of men helping men forced the unresisting prisoners into their clothing.

'How the hell do we get them to stand and walk!' demanded Manson.

It was Blair who guessed the way, rapping out the command in Vietnamese. There was immediate response and more eyeblinking from McCloud. They formed file and came forward, Page automatically putting up his hand for guidance against the shoulder of Bartel, who was the man ahead.

The Green Berets were bunched outside. Men who less than an hour earlier had swept into a barrack-room of mostly sleeping soldiers and killed them without compunction or hesitation looked among themselves, shocked at the sight of the robot-like Americans.

'Move it,' ordered Blair, continuing on.

They formed a protective file, the Green Berets grouped around the released prisoners who, to Blair's Vietnamese urging, crocodiled dutifully towards the exit and allowed themselves to be manhandled through the wire-cut entries. Only Abrahams remained, wiring the buildings behind him and moving gradually in their wake. At the main gate Blair sent Wright and Andrews to check out the guard post by the jetty. It was empty. While Lewis crossed the ignition wires on the escape boat, Andrews booby-trapped all the communication equipment in a radio shack near the larger of the two jetties and Abrahams mined the jetties and the boats they intended to leave behind. The remaining commandos carefully got the four released men into the boat: Penley, from the tower, was the last to board. He stared at the rescued men and said 'Jesus!'

'We're making a run for it,' announced Blair. 'As quietly as we can, for as long as we can. If, for any reason we're split, we're rendezvousing at the stashed boats. But I shan't wait: you've all got homing radios. If you're abandoned, fixed-time transmissions six every evening, local time, from as near to the mouth of the Delta as you can get. OK?'

There were grunts and murmurs of agreement.

'Anything in the way gets blown away,' added Blair. 'Let's go.'

The special group arranged themselves, weapons ready, around the edge of the boat and Lewis crouched over the engine. He touched the wires and there was a sluggish turn-over of the engine but it didn't catch. He primed it further, tried again and this time it almost fired. The noise seemed to echo throughout the surrounding jungle. On the third attempt it came to life, died and then started again, an uneven, untuned throb. Low-throttled, Lewis reversed the boat out, turned and steered down river purposely slowly, once more worried about noise.

Blair left the rail-edge, scooping down in front of the catatonic men, trying to reach them. Having got a response to orders in Vietnamese he tried the language to communicate but their reaction to that, beyond instruction, was as unmoving as it had been to the approach in English. He gave up after fifteen minutes, standing near Lewis at the stern and warning 'Careful of Giang island, to the left.'

Obediently the sergeant moved further to starboard, to get into the centre of the passing channel. Blair edged forward, to get alongside Manson and said, 'What do you think about light?'

'Maybe an hour,' said the major.

'That'll put us at Phung Hiep around the time the boats sail.'

'Which will be tough on the boats,' said Manson. He nodded behind and said, 'How are they?'

'Blanked out,' said the Colonel. 'Impossible to get through to them.'

Giang was about five hundred yards astern when the first explosion occurred. It was distant, a low, crumping sound but then the chain reaction began as one detonation triggered another so that they continued in a rising, belching noise and the sky far behind them was suddenly set alight in a glowing, yellow blaze.

'How's that for the fourth of July?' said Abrahams, proudly.

'Fast as you like,' Blair said, to Lewis.

The sergeant opened the throttle and the sound increased, the shuddering of the imbalanced engine vibrating through the craft: it still moved agonisingly slowly.

'They'll be able to block the river mouth against us at this speed,' said Manson.

'We're not going to try it at this speed,' said Blair. 'And they haven't found us yet.'

Behind them there was a fresh eruption of secondary explosions and a renewed glow in the lightening sky. It was Penley, in the prow of the boat and therefore furthest away from the noisy engine, who heard it first, turning into the boat and shouting 'Helicopters'.

'In,' commanded Blair.

Lewis at once swung the tiller to bring them into the bank. The overhang was meagre compared to where they had hidden their own boats but there was still some growth and when Lewis took power off they grabbed out, pulling themselves beneath its covering. Dawn was sufficiently close for them to see the three machines outlined black against the sky, chopping the air as they followed the river line up towards Can Tho.

'They'll be checking what happened,' judged Blair. 'It's not a search, not yet.' The Colonel allowed the machines barely time to pass and then said to Lewis, 'Let's keep going.'

The sergeant brought the boat out again but kept closer to the bank this time, in readiness. It was fortunate because within minutes Penley, still at the stem, called softly, 'River boats'. Lewis thrust in towards the bank again, killing the engine so that they drifted into the protection of the tree-line. This time only two men, Wright and Abrahams, clung against the overhanging branches: the rest formed silently, against the starboard side, waiting. The lead boat was a military cutter, with a gun, maybe 76 mm, front-mounted, and behind came a smaller troop carrying craft. They were moving quite openly, clearly not suspecting anything else on the river; their gossiping, shrill voices easily reached the tensed Americans. This time Blair waited until the Vietnamese had rounded a curve in the river before nodding to Lewis to start their own engine: warm now, it came to life at once.

'We couldn't outrun that cutter, not even in our own boats,' said Manson.

'Let's hope we don't have to try,' said Blair.

'It's going to be near light at Phung Hiep,' said Manson.

'And definitely light at Long Phu,' said Blair.

They edged towards the first township hugging the far bank, trying to use whatever shadowed blackness remained in the night, able to throttle right back with the current behind them. The far side was darkened too, so that it was difficult to discern any movement among the piers and jetties, cankered with fishing boats. The pink blush of dawn was already coming to the sky and far back, appearing beyond the jungle now because of the twists in the river, the different glow from the explosions created another, harsher colour.

'The noise probably reached here,' said Manson softly. 'Woke people up.'

'Let's hope they're looking towards the blaze,' said the Colonel.

They were past now, going into the bend that would take them out of sight of Phung Hiep. To hasten their concealment Lewis took the boat away from the bank, steering into mid-channel to get more quickly behind the out-thrust bump of land. It put the boat directly in the centre of the river and in full view of four fishing vessels that sailed from Phung Hiep minutes before they began the avoiding manoeuvre. The Vietnamese craft were about five hundred yards ahead, being oared-steered with the following current: the bird-song talk of the previous morning stilled as the two groups stared at each other.

'Close up,' ordered Blair. 'We want them all.'

Lewis increased power, reducing the gap and Manson waited until they were less than twenty yards away before he said 'Fire'.

The Vietnamese appeared to realise what was going to

happen at the moment the first shots ripped out, screaming and trying to hide behind the fragile gunwales of their boats. One man tried to throw himself overboard and was hit in mid-air twisted grotesquely out of the arc of his dive by the impact of the bullets, and another was offering his hands in a praying gesture when he was struck and hurled back against the masthead. It was over very quickly, less than two minutes: an explosion of shooting and then abrupt, echoing silence. Uncontrolled, the boats twisted and turned in the eddying currents.

'Destroy them,' said Blair.

Penley and Abrahams threw the grenades, lobbing them easily into the boats which burst apart under the detonation. A scum of floating wood, canvas and netting spread out over the yellow surface of the river: a body was trapped, in one of the seine nets, only the hand jutting upwards through the surface as if seeking help.

'They'll know soon enough now,' said Blair. 'Let's get to the boats.'

Lewis wound the throttle up as high as it would go, careless of the noise or the juddering protest from the engine, sweeping from curve to curve in the quickest route, risking mudbanks to maintain the shortest, most direct line. They swept past one apparently still slumbering hamlet, everyone looking out now for the recognisable overhang beneath which their faster craft were hidden. Penley, from his superior vantage point, saw it first, giving the indication to Lewis at the moment when they all became aware of the whock-whock sound of a helicopter, lower than those they had seen earlier. Lewis drove full throttle into the cover, reversing at the last moment to lessen the impact but still jarring into the soft bank mud. The Green Berets were ready for the impact but they had forgotten the rescued men. The four hurtled forward, unprepared and unprotected, blood gushing immediately from a gash in Page's head where he collided with a rowlock. He showed no distress or pain at the injury. Abrahams attended him from his own medical pack, while Penley and Wright tried to straighten the others. There was a hole, where the boat had burst into the cavern and Blair tensed against it being recognised by the searching helicopter. Sure of the concealment of the overhanging darkness, he stared out openly, watching the machine. It skittered and swept twenty-five feet above the water, shivering the surface with its downdraft, obviously searching. Blair looked sideways, at movement to his left, and saw Manson in position, ready with a grenade launcher.

'Andrews, Wright,' he called. The two men were immediately alongside. 'Rapid fire,' he ordered.

They took up their positions either side of the major, their M-16 on rapid fire. Briefly Blair thought the observer had missed the entry point. The helicopter appeared to be continuing downstream and then suddenly it jerked right, coming directly towards them, settling lower against the water for the crew to look into the darkened, torn hole.

The rotor draught tore at the leaves, stripping back their cover and Blair said 'Now!'

The cockpit fissured and burst under the blast from the two guns, caving in at the impact of Manson's grenade which tore the helicopter apart in an eruption of orange and black. It turned completely over and landed rotors first, hissing, into the water and settled beneath the surface in a foam of bubbles.

Behind them the rubber boats had been brought down and the second one inflated and stressed with its centre-planking. They transferred the engine to the back-up boat, tested the ignition of both and ensured the fuel level of both. They divided the rescued men, two to each boat, while Blair composed his message to the waiting freighter. It was a radio specifically designed for clandestine communication, capable of 'spurt' transmission to relay in seconds a message occupying minutes of normal speech. Blair reported fully on the condition of the men, their escape from the camp and progress so far down the river and his intention to make a dash the remainder of the distance, asking the freighter to be in position and for the aircraft carrier to provide air cover if required. He replayed it, to ensure he'd forgotten nothing, then depressed the relay button. The message went in five seconds. Manson was in the secondary craft so Blair got into the other, in which he'd made the original incursion.

'They'll be waiting for us,' he promised, generally. He hesitated, then said 'Go like hell!'

He burst first through the already created hole, arcing the easily manoeuvred boat in a sweep to take him into the middle of the channel, opening the throttle wide when he got to the deep water. Warned because of what had happened earlier, Penley and Wright held tightly to their passengers, McCloud and Page. The others knelt as before, arms looped through side-halter ropes to steady them in the bucking boat. Manson was immediately in their wake, in perfect convoy. They flashed by the awakening Long Phu and almost immediately afterwards came upon the fishing fleet, spread out like an amateur regatta. This time there was no thought of attack, only passage through. Blair chose the widest gap, bouncing through, the combined wake jostling and rocking the amazed fishermen.

The dividing islands of the river-mouth formed in front, like whales gradually rising from the water and Blair squinted in the early morning brightness, seeking any block. He was practically into the channel through which they passed originally by Cua Tranh De when he saw the military boat edging out and at the last minute swung into the middle waterway, putting the tiny island between themselves and any pursuers.

They crashed into the waves of the open sea, engine howling as the swell momentarily lifted the propellers completely clear of the water.

From behind Lewis shouted 'It's a cutter: it's gaining.'

'Second boat?' Blair yelled back.

'Keeping pace,' reported the sergeant. 'Cutter's closing to about fifty metres. Rapidly.'

From behind, over the engine noise, came the stutter of machine-gun fire and the louder, thudding sound as Abrahams, in the second boat, tried to respond with the grenade launcher. It fell hopelessly short. Andrews alongside the man, sprayed back with his M-16, not with any particular aim, wanting only to keep down the marksmen in the following boat.

Hawkins, Jordan and Patton were on the bridge of the freighter, straining between the horizon for the first visual sighting and the radar screen. It was on the screen that they detected movement first not from the rubber boats, which defeated any reflection, but from the metal-hulled cutter.

'Seven miles,' said Patton. 'It's too far!'

'What about the air cover?' demanded Hawkins, eyes narrowed against the glint from the sea.

'Seeking Washington's approval,' said Jordan. He was beside the Englishman, eyes shielded beneath his cupped hand.

Out to sea Blair was shouting his instructions to the other men in the boat. Lewis helped Penley and Wright rope their flopping passengers against the wooden deck stays, to keep them inboard and then Lewis semaphored in their signal language to Manson's boat, so the major would respond. At the answering signal that the other two men were secured, there was a flurry of hand and arm movement from the sergeant. Abruptly, both rubber boats reduced power, so that the pursuing Vietnamese cutter bore down upon them. And then both peeled away in opposing directions, Blair to starboard, Manson to port, in the tight, hard circles of which the boats were capable. The effect was to take them both from the direct firing line of the cutter and at the same time place the Vietnamese ship between their crossfire. Blair and Manson completed the circle, accelerating to run parallel with the ship while their men pumped in machine gun fire from either side and this time scored direct hits with grenade launchers. There was the thump of an explosion, a billow of blackened smoke and the cutter slumped into the water, all power gone as its engines were destroyed. Without pause Blair increased speed again, picking up the lead position and driving out further into the open sea.

'Whatever it was it's stopped,' reported Jordan, at the sophisticated radar of the supposed freighter. 'It's dead in the water.'

The other two men came beside him, staring down into the green screen and Hawkins pointed and said, 'What's that?'

'Aircraft,' said Patton, quietly.

It was Lewis who saw them, in Blair's boat, shouting the warning to his Commanding Officer. 'Helicopters,' he yelled. 'Two as far as I can make out.'

At Blair's instruction Lewis gestured backwards again and at once the second craft followed the example of the first and began using its manoeuvrability, twitching back and forth, always avoiding a straight line. They were Russian helicopters, Tupalov, with pod-mounted cannon and two sets of underslung rockets, four on each side. Each machine fired, hopelessly wide and their cannon, too, churned harmlessly into the water. From the wildly jerking rubber boats the firing of the Green Berets was equally ineffective.

'Three miles,' counted off Patton, on the freighter. 'Where's the carrier planes, for Christ's sake!'

At sea the helicopters attempted a combined tactic, one squatting low in an effort to swamp the skittering boats in the force of its downdraft and create a stable target for the other. It might have worked upon less well-trained men, panicking them with the deafening noise and the turmoil of water. Instead it presented Lewis with an easy target. He was still careful with it, leaning out to avoid wasting bullets on the metalwork, aiming for the reinforced bubble of the actual cabin. It shattered and instinctively the pilot veered sideways, presenting an even easier shot. Penley joined the sergeant, both firing into the cabin. The bubble burst inwards and they were close enough to see the pilot jerked up against his restraining straps by the impact of the bullets. Uncontrolled the helicopter pitched sideways and fell into the sea.

'There!' said Jordan, on the bridge of the freighter. They could see the helicopter more easily, blackly outlined against the whitish-blue of the morning sky and then, through glasses, located the boats. Gradually the battle formed in front of them, an unwanted tableau.

'That helicopter won't stop at any international limit,' guessed Patton. 'And we haven't got any armament on this ship.'

'We won't need them,' said Hawkins. 'Look.'

Fighters from the unseen aircraft carrier were swarming in from the east, low against the water in tight formation. In the boats, Blair was the first to see them, shouting and gesturing first to Lewis and then to the following Manson. The crew of the helicopter saw them too, pulling up high and firing the last of its rockets in a flurry, without aim, hoping for a lucky shot. There wasn't one.

The misjudgment was Blair's. It was his first and it was fatal. He'd seen the machine jerk upwards and loose the wasted rockets and imagined it was pulling away, in face of the fighters, which it was but not immediately. Blair straightened the craft, abandoning the manoeuvring, giving the helicopter the target it had sought since the pursuit from the mainland. There was an abrupt stutter of firing as the machine banked, on its homeward return. Blair caught the full blast, thrown forward in the boat with his back pulped: Penley was hit high in the left arm and almost hurled out of the suddenly unsteered boat. Lewis grabbed the tiller arm, bringing it back on course and Wright crouched over Blair, who was quite dead.

Over their heads there was the scream of the arriving jets and the slightly louder hissing sound of the rockets being fired. The helicopter caught the first salvo, bursting apart in a spray of exploding metal.

Lewis didn't reduce speed. Out of danger now he maintained the direct line, Manson close behind. Wright crouched in the boat, cradling the Colonel's body, not knowing what else to do.

The sea was calm, without any lift, but there was still a drop of almost twenty feet between the boats and the freighter rail.

'We'll need help,' shouted Lewis.

Scramble-nets were thrown over and crewmen climbed down, shouting back for hoists. The derrick was swung over and both boats hooked fore and aft and lifted bodily from the water, everyone remaining inboard.

Hawkins, Jordan, Patton and the psychiatrist, Hamilton, were on the deck when they were swung in and lowered to the deck. Wright, in the first boat, released the protectively secured prisoners: Abrahams did it in the second craft. The four remained crouched where they were until Lewis told them, in Vietnamese, to stand. They did so, unsteadily, holding out to each other for support, like disorientated children.

'Dear God!' said Patton.

'I'd better tell Washington,' said Jordan.

The President was too excited to sit. He walked among the assembled Task Force, driving the fist of his right hand into the palm of his left, saying 'We've done it! We've got them out! Can you imagine that! We've got American prisoners out of Vietnam!'

'I don't like the early indications about their condition,' said Milton Snow, warningly.

'They're bound to be shot, at this stage,' dismissed Harriman, refusing any curb on his euphoria. 'Imagine what they've been through!'

'We lost Colonel Blair,' reminded General Bell.

Harriman stopped pacing, looking directly at the Army Chief of Staff. 'The Congressional Medal of Honour is the highest I can personally award,' said the President. 'I'll make the declaration at the same time as I announce the rescue . . .' He hesitated, mind occupied with achieving the maximum electoral impact. 'Prime time announcement, of course,' he said, more in private conversation with himself than with the others in the room. 'I'll send Air Force One to the West Coast, to bring them here to Washington. Reception on the White House lawn . . .' Another pause. 'Medals for them, too. Congressional as well . . .' Harriman stopped, his mind occupied with every eventuality.

'We'll time the announcement carefully,' he decided. 'Make the demand through the United Nations and every other world body for Vietnam to be accountable for other Americans still held. Use it to get more concessions from the Russians, in Geneva, too . . .'

Harriman smiled around at the assembled men, the majority of whom depended upon his remaining in office to retain their own jobs. 'I would think, gentlemen, that a second term is guaranteed, wouldn't you?'

Chapter Twenty-Five

There was neither excitement nor exhilaration, not even when the President's congratulatory cable was followed by a personal telephone call patched through from the White House and relayed over the freighter's echoing public address system, announcing the posthumous award to Blair, decorations for the rest of the squad and invitations to them all to the Washington ceremony of welcome for the freed prisoners.

Instead men gathered in subdued, barely speaking groups: overhead the carrier fighters flew close escort, the freighter throbbing at maximum speed in the direction of the still unseen warship. Hawkins searched for his own attitude and decided – first surprised, then irritated – that it was embarrassment, the embarrassment of a stranger intruding into other people's tragedies. Which was ridiculous, he recognised, the irritation increasing. They'd freed the men they set out to rescue, the need for secrecy no longer existed and he had a story to better any in which his father had ever been involved. And more. The opportunity, once Forest and Bartel and McCloud emerged from their initial examination in the wardroom above, maybe to find out what had happened at Chau Phu: the disorientation would have gone, once they realised they were free.

As unprofessional as he knew it to be, the feeling of intrusion – together with Blair's earlier rejections – made him initially hesitant to approach the Green Berets but at last he did, needing to learn about the rescue. There was a reluctance at first, the anticipated rejection, but gradually their responses changed, led by Manson and almost at once supported by Lewis: Hawkins became aware they actually wanted to talk about the incursion, as a catharsis for their feeling at Blair's death, at the very moment of success, and as a tribute too, to the man's leadership. Hawkins prompted, occasionally, to achieve some direction to the story but for long periods he was able to let the tape run uninterrupted: there was so much detail, finally, that he twice changed cassettes, increasingly curious at the repeated references to the men's mental numbness.

Hawkins had just completed the interviews when there was a sound different from the overlaying noise of the escort fighters. He emerged on to the mid-deck companionway in time to see the transfer helicopter arrive from the carrier and settle itself neatly into the stern of the freighter and he realised he had a time limit. Hawkins went quickly back into the ship, using the inner corridor to reach the wardroom. He arrived at the same time as Jordan and Patton, descending from the bridge.

The CIA man was coolly controlled as always but Patton was suffering badly from the heat, familiar perspiration bubbled on his forehead and upper lip. He twitched one of his uncertain smiles at Hawkins and said, 'We're going to be shuttled to the carrier; I presumed you'd want to go.'

'Thanks,' said Hawkins. He hoped there would be a sufficient interval for him to write all that he had; and that the communications from the carrier were as good as they appeared to be from the specialised *Elmer C. Gorst.*

Jordan knocked and at Hamilton's shouted permission led the way into the room. The psychiatrist had put the freed Americans side by side on the seat that ran the interior of the bulkhead and set a chair for himself directly in front of them; as he went into the room and looked at the men Hawkins had the fleeting impression of inanimate window dummies waiting to be dressed for some display.

'The helicopter's arrived,' Patton said unnecessarily, to the psychiatrist.

'Ready to go?' asked Jordan.

'The sooner the better,' said Hamilton, turning to them. 'They're in bad shape: far worse than I feared when I saw them down there on the deck.'

'How bad?' said Patton.

The psychiatrist shook his head. 'Impossible to say with any accuracy, from what I've tried to do here. But I've never seen men like it. It's not just indoctrination: they've been pumped full of God knows what drugs and because we don't *know* we won't be able to counteract it.'

'But they will get better?' said Jordan, seeking assurance.

Hamilton shook his head. 'I'd like to think so but I wouldn't make any forecasts from what I've seen so far.'

'What's *happened* to them?' said Hawkins, entering into the conversation.

'I don't know,' replied the psychiatrist, honestly. 'Like I said, I've never seen men like it. I think their minds have been emptied: either that or they've erected their own mental barriers and won't come out from behind them. They're not showing the slightest awareness of where they are. Of being rescued. I'd expect them to be traumatised to a degree by the very fact of their rescue but it hasn't registered at all. They've no comprehension of being back among Americans, in American safety.'

'I thought the woman was exaggerating when she spoke of the mental pressure,' said Jordan.

'She wasn't exaggerating enough,' said Hamilton. 'These men are going to need specialised care for a very long time.'

'Then let's get them to it,' said Jordan, abruptly.

Lewis, whose Vietnamese had been necessary to get them into the wardroom, was summoned again to get them out. They filed dutifully one behind the other, Page's hand coming out automatically for the guiding shoulder of Bartel in front. The Green Berets and crewmen stood silently watching as the four emerged on to the stern deck, towards the helicopter. Forest was leading and he was about ten metres away when he suddenly baulked, pulling back with a whimpering, mewing sound: the three behind jostled into him and then Bartel appeared to become aware of the droop-winged machine, grunting some unintelligible sound.

Lewis started to speak, in Vietnamese, but Hamilton shouted 'No, leave it!'

The psychiatrist hurried up to be in front of the ex-prisoners, staring intently into their faces. To Lewis, the nearest Green Beret, Hamilton said, 'How many helicopters chased you, from the moment you freed them?'

'Three,' said the sergeant, at once. 'We shot one down in the river: the other two you know about.'

Hamilton nodded. 'Enough for a reaction! At least something's getting through to them.'

'Is it important?' demanded Patton.

'I can't tell, not at this stage,' said the psychiatrist. 'Certainly I don't think we should risk any further trauma by forcing them into a helicopter. I think we should make a sea rendezvous with the carrier.'

'That'll take several hours,' warned Patton.

'I don't care how long it takes,' said Hamilton.

Obedient to Lewis' fresh instruction the four Americans turned and filed back into the ship and the wardroom.

'How many hours, until we get to the carrier?' Hawkins asked Patton.

'Eight at least,' said the ship owner. 'I suppose you could use the helicopter if you wanted to fly there.'

'I'll stay here,' said Hawkins. 'I've got a lot to do: I can use the time.'

He was, in fact, to write uninterrupted for only five hours.

The warning of an intended but unspecified announcement was regarded sufficiently newsworthy to make the midday newscasts, on both radio and television, and at 5 pm Peterson fronted his own hastily compiled trailer, suddenly inserted into the programming and urged television viewers to tune in. Forewarned, the President was in his private working suite adjoining the Oval Office, with his immediate campaign advisers, when Peterson appeared on the screen, on prime-time coast to coast hook-up.

Peterson looked solemn, in dark tie and suit, flopping hair for once lacquered into place. 'I have taken this time, a time when I hope the majority of Americans can be reached, to speak to you tonight upon an event of major importance . . . of historic importance . . .' he began, grave-voiced.

'What the hell is it?' demanded Harriman.

'Eight years ago I was involved in an episode in Vietnam in which Americans were thought to have died. Thought, among others, by me to have died: I testified to that at a later enquiry . . .'

'What in the name of Christ . . .' shouted Harriman, half-rising from his chair.

'I have today learned that I was wrong. Less than twenty-four hours ago three men whom I mistakenly believed to have been killed were rescued from Vietnamese imprisonment and tonight are on their way back to safety and comfort in the United States of America . . .'

'The son of a bitch is making it sound as if he's entirely responsible: as if he did it!' screamed the puce-faced President. Then he said, 'The Englishman! It's got to be that bastard Englishman!'

The outraged Washington cable, demanding a full explanation from Jordan, arrived thirty minutes ahead of a similarly outraged, explanation-demanding message from Hawkins' London office.

Chapter Twenty-Six

Peterson followed his television address with a full press conference at which he announced he was making available his campaign aircraft to fly the wives of the freed men to the West Coast and issuing a challenge to the Soviet Union – 'the true controllers of Vietnam' he called them – to force a declaration from Hanoi about other POWs still held. Under questioning he repeated the acknowledgment of his mistake – 'anyone can be wrong: it is an error for which I will always feel guilt and regret' – in believing that Forest, Bartel and McCloud had died in the ambush and by what he did not say, rather than what he did, allowed the impression that their rescue was in some way his atonement.

Harriman's campaign staff were panicked into a response and it looked panicked. The President's television announcement and press conference followed Peterson's by several hours, giving the appearance of the man trying to catch up and in his outraged anger – an anger worsened by the obvious lack of confidence among his campaign managers – Harriman badly prepared himself on what rescue details had been relayed from the freighter and gave a hesitant, uncertain performance which contrasted badly with the senator's easy, controlled manner.

Peterson's demand to Moscow completely pre-empted the stance Harriman intended to make and his repetition of it sounded like the echo of a good idea his political rival arrived at ahead of him. Even the announcement of the valour awards sounded flat because moments before Harriman got to the briefing podium the message arrived from Jordan warning that the men's mental condition made such a public reception at the

White House unwise. It meant Harriman going instead to the West Coast, something else Peterson had already told the country he intended doing.

Afterwards Harriman was impatient with his election staff, not needing to be told what a bad showing he made. Instead he called a meeting of the Task Force – whom he insisted remain in the White House anyway – picking out the CIA Director for attack.

'What's the answer?' Harriman demanded. 'How did it happen?'

'We haven't heard back yet,' apologised Milton Snow.

It was not to take long.

In an enclosed, CIA-controlled vessel like the *Elmer C. Gorst*, with an inviolable insistence upon records, the source was located simply by checking the radio log. Patton's reaction, when Jordan and Hawkins confronted him in the displaced Captain's cabin, swung from stuttering defensiveness to open belligerence.

'Why!' demanded Hawkins, incredulous as well as angry, still trying fully to absorb what he'd lost. 'Why the hell did you tell Peterson!'

'This is my ship: I can do what I like in it,' fought back Patton. There were black sweat-marks staining his shirt and a nerve was pulling just below his left eye.

'No,' rejected Jordan at once. 'This ship is wholly leased to and operated by the Agency. And runs under its jurisdiction. You knew that. Just as you knew the need for secrecy.'

'*Before* the rescue,' insisted Patton, in desperate qualification. Gesturing to Hawkins, he said, 'He was going to write about it!'

'Because there was an agreement: an agreement he stuck to,' said Jordan.

'And which you've fucked,' said Hawkins. There was a fresh surge of anger at his impotence to reverse what had happened. 'Do you know what you've done!' he said.

Patton's demeanour abruptly changed, back to defensiveness. 'I'm sorry,' he said. 'I'm really sorry. I had no idea that Peterson would do what he did: that he'd make it public like this.'

'But *why* did you tell him!' repeated Hawkins. He wanted to hit the man, physically to hurt him. 'You had no *reason*.'

'I thought he should know: he was involved, after all. I just thought he should know,' said Patton, lamely. The eye-flicker worsened.

'Crap,' said Jordan. 'What's the deal, between you and Peterson?'

'There isn't any deal,' said Patton.

Jordan waved the Washington cable. 'This is from the White House,' he said. He paused, surprised at Patton's lack of reaction. 'The *White House*,' he said again. 'They want an explanation.'

'I made a mistake, OK?' said Patton.

'No,' refused Jordan. 'It's not OK. What was the deal?'

Patton looked away, not immediately replying. 'Peterson helped me, in the beginning,' he disclosed, soft-voiced. 'I needed set-up money and he gave it to me: a lot of money and he didn't press me for repayment, like a bank would have done. I paid it all off, a long time ago. But I figured I owed him . . .' He looked up, the belligerence returning. 'I *did* owe him. I'm a supporter, politically. And I thought that politically what has happened might embarrass him . . .' There was another pause. 'But I never expected him to go public like he did. Honestly I didn't.'

'Jesus!' exploded Jordan, at the explanation. 'I wouldn't like to be you, buddy. I really wouldn't like to be you.'

Hawkins, still wanting to hurt the man, said, 'Was this the way you wanted Sharon Bartel to find out!'

The question brought the most positive response from Patton since the argument began. He felt up to his eye, trying to stop the vibration, pulling his lower lip between his teeth. 'No,' he said, quiet-voiced. 'This isn't how I wanted her to find out.'

'I didn't make any promises,' said Hawkins. He wasn't feeling any satisfaction from the bullying. Jordan looked curiously between the two men.

'I guess you've got the right,' said Patton, defeatedly.

'What's this about?' said Jordan. 'And this time I want to know.'

Patton told the other American about Sharon Bartel, a stumbling explanation, and Jordan shook his head, face twisted. Then he said, 'You're right: you've got a lot of difficulties.'

'We all have,' said Hawkins. 'Too many to waste any more time here.'

He delayed responding to the London cable until he wrote the account of Ninh and Nicole's escape – already half prepared in note form – the location of the camp and the rescue of the Americans. He opened the telex line with the explanation and let the articles start immediately afterwards, using the transmission time to complete the rescue story, maintaining a running file between the ship and England. There was so much material that its sending occupied the entire time it took the freighter to rendezvous with the aircraft carrier. They transferred by cutter, the freed Americans and the psychiatrist in the first, the unspeaking Hawkins, Jordan and Patton in the second and the Green Beret squad in the third.

There were messages already waiting, when they got to the carrier Captain's cabin. London merely acknowledged receipt of Hawkins' file, without any comment, adding only 'Jones backstopping fullest from America', which furthered Hawkins' depression. Patton stared down at his slip and said to Jordan 'They want me in Washington, right away.'

'Are you surprised?' said Jordan, unsympathetically.

An executive officer whose name Hawkins missed said to Patton, 'We've got instructions, too, sir. There's a plane waiting, to take you on to Guam.'

Patton started following the officer from the room and Hawkins called after him, 'I didn't do it.'

The ship owner turned, frowing.

'Not because of you,' said Hawkins. 'I couldn't give a damn about you. I didn't write it because I think Bartel has got enough problems. His wife, too.'

'Thank you anyway,' said Patton, emptily. 'For Sharon that is.'

'I admire you,' said Jordan, after the other American had left the cabin. 'If it had been me, I'd have screwed the bastard every way possible.'

'It wouldn't have been him I was screwing, would it?' said Hawkins.

It was two hours before Hamilton entered the cabin. There were three other white-smocked men with him: he didn't bother with introductions.

'We're wasting our time,' declared the psychiatrist. 'We've

got to get them home and into some sort of intensive care.'

'This isn't turning out quite to be the sort of triumph everyone expected, is it?' said Jordan.

In Washington the furious President was reaching the same conclusion. To the CIA Director he yelled, 'Why didn't we know about Peterson financing the fucking man, in the beginning! He's been working for you for years! Didn't it come up on the security clearance check!'

'No,' admitted Milton Snow, aware of the collapsing relationship between himself and the President. 'It should have done but it didn't. Seems it wasn't anything official, with bank records at first. Patton paid it back in full, $100,000, in instalments over a year.'

'I want Patton,' said Harriman, determinedly. 'For what he's done to me I want him twisting slowly, from a hook. I want everyone to understand that.'

In Guam, with its complete medical facilities, the former prisoners were bathed and shaved and dressed in new uniforms, physically to improve their appearance, and for the flight across the Pacific each travelled accompanied by two medical orderlies, with doctors in attendance, too. Sharon Bartel and the other wives had accepted Peterson's quick invitation, which meant the President had to allow the senator on to the same welcoming platform with other senior members of Congress. Despite Jordan's repeated warnings, no one had properly appreciated the condition of the men and it was Peterson's arms the television cameras showed around the shoulders of the sobbing Sharon Bartel and the shocked wife of McCloud, because he was nearer to the women when they stumbled, both near to collapse.

The rattled President made a misjudgment, launching into a speech carefully prepared to rebalance his initial losses, realising too late how inappropriate it was with television cameras focussed on men shown to be little more than mindless shells. He brought it to a confused, embarrassed conclusion, abandoning at the same time the presentation of the intended decorations, to which he'd already referred, worsening his performance at the ceremony in view of the whole country.

Throughout Hawkins strained towards the official party, looking for Eleanor. He decided, regretfully because he was anxious just to see her, that she wasn't there, balancing the regret with his acknowledgement of how much he hated the thought of her physically being anywhere with Peterson anyway.

Because of his presence aboard the freighter he had been given a favoured position at the welcome, close to the dais. After the numbed men had been guided to ambulances and the President escorted away and their wives comforted from the platform, Hawkins shouldered his way towards it, determined to confront the senator.

Rampallie saw him approaching first, leaning towards Peterson, so the politician had turned towards him by the time Hawkins reached the dais.

'Ray!' said Peterson, as friendly as always. 'Where have you been!'

'Working, aboard the freighter. And having the ass beaten off me.'

The senator's smile faded, at the obvious anger. 'What the hell's wrong!'

Instead of replying Hawkins looked to Rampallie and then back to Peterson. He said, 'I was told that you'd gone as high as the Secretary of State to get Nguyen Ninh out of Vietnam?'

Peterson frowned towards his campaign manager, as if he had difficulty in remembering the details. 'That right, Joe?'

'Yes, sir,' said Rampallie, at once.

'State say there was never any approaching. Nothing at all.'

'We tried, Ray. I tell you we tried.'

'Like you tried by making the announcement about the rescue!'

'I didn't know you were involved, Ray. You didn't tell me. I *didn't* break it, not the moment I heard. I tried to call you. I called you three times and left messages on your answering service. You can't disappear for weeks, without even a "hello" or a "goodbye" and expect me to do more than that, can you?'

On the platform the people who remained were shifting uncomfortably, aware of the argument. The attentive Rampallie felt out for Peterson's arm, to move him away from a public dispute but Peterson refused to leave at once. 'Let's talk again, Ray. Not here but let's talk again. Let's not foul something up with a misunderstanding.'

Hawkins was left on the dais, feeling embarrassed and

ineffectual. He knew he was right to have demanded an explanation but equally he knew that his irritation at the senator was fuelled as much by what was happening between himself and Eleanor. He went immediately to the airport, anxious despite his fatigue to get back to Washington. He found it difficult to sleep on the aircraft, although he was tired. Once, he reflected, he would have sought help in booze. Now he preferred to remain half-awake. Overtired by the time he landed at Dulles, Hawkins went directly to the office, the idea having occurred during the flight.

Harry Jones had taken the messages during his absence but the answering service kept records and it was easy enough to check the truth of what Peterson had claimed, in California. There *had* been three calls, two of them personally from Peterson. To be absolutely certain, Hawkins checked both with ABC and NBC, confirming that all three were timed before Peterson's public declaration.

Hawkins knew he would have to apologise and was trying to decide how when the telephone rang and Pearlman said, 'I'm glad I caught you: tried the house first.'

'What is it?' said Hawkins. Although he felt mentally alert his body ached.

'Had the Peterson people on to us at State, creating a hell of a row about Nguyen Ninh. Seems I misled you that day in the ambassador's office, saying there hadn't been any approach for help. There had been. Katzbach tried in the wrong direction, to the Secretary himself. Apparently it came in at low level and some half-assed assistant didn't realise it was from a senator and filed it under "No Action".

'Oh Christ!' said Hawkins, despairingly.

'Hope we haven't caused too many problems.'

'Forget it,' said Hawkins, confronted with two apologies.

He was slumped, half-dozing at his desk, when Jones entered, surprised to see him and halting in the doorway.

'Nothing to celebrate, surely!' demanded Jones, misunderstanding. 'Or is it drowning sorrows?'

Hawkins tried to blink into full wakefulness. 'I'm very tired,' he said. Why the hell did he have to explain to this man?

'I'm surprised,' said Jones, coming further into the room. 'From here and from London it looked as if you were asleep on the job.'

Hawkins hit him. It was a clumsy, awkward effort, Hawkins realising what he was doing at the last moment and opening his fist to lessen the hurt, so that it was more of a slap than a blow and the more alert Jones ducking away, so that he was only struck on the side of the head, with hardly any force. It was still sufficient to dislodge his glasses, which clattered on to the desk. The left lens shattered.

'What do you think you're doing!' demanded Jones, high-voiced from outrage and nervousness.

Hawkins collapsed back into the chair, further embarrassed. Another apology, he recognised. Just as he recognised that he had tried to take out against the man blinking myopically at him across the desk his frustration at Eleanor and the freighter and Peterson.

'I'm sorry,' he mumbled. 'Tired. Very sorry.'

'You're going to be sorry!' said Jones, aware that the danger had passed and wanting to fight back, although not physically. 'I said it was going to be a disaster with you involved and it has been. They're going to promote it as a major series because they've got to recover and they haven't any option but everyone knows we've been beaten: knows what a disaster it's been.'

Ironically – obscenely – Jones was wrong.

Everything would have been lost if the rescued men had been able to give press conferences but they weren't. The California return fascinated as well as shocked not just America but the countries in which the television pictures were relayed, which was a huge number throughout the world. The effect was to create a demand, not disinterest, in what Hawkins wrote.

Two weeks after the return, syndication arrangements had been made with over two thousand newspapers and magazines – five hundred of them in America – and their own newspaper, which naturally insisted upon publishing first, registered a circulation increase of 60,000 copies on the first Sunday edition in which it was carried, rising to 70,000 on the second. Illustrating the complete reversals which newspapers – like most businesses – are capable, the belated congratulatory cables arrived from the editor and proprietor during the third week, with the recall message from Lord Doondale.

There was a personal letter from Peterson, dismissing Hawkins' written apology as something already forgotten and asking when he was going to join the campaign. Hawkins, not wanting to accompany the senator when Eleanor was with him, used the return to London as an excuse for immediately not accepting but promising to join in the future.

Going back to England meant, too, that Hawkins missed the convention at which Nelson Harriman was adopted as the continuing candidate by the narrowest margin in fifty years of his party's history.

Chapter Twenty-Seven

It was an English summer's day, softly warm, not brutally hot like the Washington he'd left the previous day and Hawkins decided to walk into Fleet Street. The same hotel and the same route as before, past the church and the law courts, he recalled. Except that before he'd thought he was on his way to be told his Washington posting was over and now he was on his way to be fêted. Hawkins knew he should have been pleased – relieved even – by the way everything had worked out but he wasn't, because he was reluctant to quit Washington, even though Eleanor had been travelling for the past two weeks with Peterson and it was doubtful if there would have been an opportunity for them to have met anyway. Any more than there had been since his return from the rescue: just snatched, hurried telephone conversations in which she said she loved him and he said he loved her and which made their separation worse rather than better.

Hawkins had gone beyond the self-recrimination, telling himself of the stupidity of it all: the danger, too. At school, privately trying to come to terms with being the son of a famous father, Hawkins had gone through a period of reading know-and-improve yourself books, written by people who had supposedly made a success of their lives. He'd lost most of the facile amateur psychology but remembered the axiom from one, always to treat every day as a separate entity, to be filled and enjoyed to its fullest. That was how he thought of himself and Eleanor, boxed and secure against anything outside, taking every moment together as the bonus it was and refusing to worry about whether there would be another or the consequence if there was. Thank God her travelling ended in a week.

Hawkins entered Fleet Street on the opposite side of the road from El Vino and halted dutifully at the traffic lights, gazing across at the bar. Something else that was different from last time, he thought. The need had completely gone, to the point where he didn't even think of booze, and if he did, not as something about which to be frightened. He decided not to drink so readily when he was with Eleanor: then maybe she'd cut down. She was too beautiful to risk damaging herself with alcohol.

He was recognised at once by the commissionaire and escorted to the private lift which went non-stop to the chairman's floor. The boardroom, where the luncheon was to be held, was part of the original building, like the editor's office on the floor below, panelled in dark wood with an enclosed bookcase containing bound editions of the newspaper from its first publication and busts of Lord Doondale's grandfather – the original founder – and father on short marble plinths. Doondale was a dried-out stick of a man who rigidly maintained his family's Quaker tradition, drank only mineral water and in whose presence Wilsher's normally flamboyant attitude was noticeably subdued.

Because there were only three of them they ate at a small round table, Doondale dominating the conversation. The circulation increase during the Vietnam rescue series had peaked at 100,000, he said, and the circulation department reported that 40,000 of those were holding. The syndication income still had finally to be calculated but in advance of the figure being settled, Doondale was awarding him a bonus of £10,000. The sculpture of Hawkins' father was well advanced and the Prime Minister had agreed to perform the unveiling, for which, naturally, Hawkins would be expected once more to come back from Washington.

'Will the book be finished by then?' asked Doondale. He had a grating, dry voice, matching his appearance, and the tendency to clip his sentences.

'I would expect so,' said Hawkins. 'It was largely completed before the rescue. I'm rebuilding it now, inserting all the Vietnam material in the front.'

'Our syndication department are acting as agents,' Wilsher said, to the proprietor. 'The interest is enormous. There are already agreed and signed contracts with America and most of

Europe. The initial print here in England is to be fifty thousand and a book club have taken a hundred thousand.'

'Want you to know I'm very proud to have you with the paper,' said Doondale. 'As great an asset as your father ever was.'

'Thank you, sir,' said Hawkins. It was, he supposed, the accolade for which he'd always striven: he expected to feel a greater satisfaction than he did.

'Anything you want?' asked Doondale. 'Like to treat our star people well, you know.'

Hawkins hesitated at the invitation. He had come expecting a celebratory lunch, not the proffered influence and he was unsure how to use it. 'I'm very happy at the moment,' he said. 'I don't really want to change yet.'

'Stay there as long as you like,' said Doondale. 'But when you feel like something different, let Wilsher know. Your father didn't stay long in one place and neither should you: don't see why we shouldn't promote you to be as great a name as he was; greater even.'

Everything he'd ever wanted, thought Hawkins; more in fact. Maybe the excitement would come when he had time fully to consider what he'd achieved. There was one thing he could try, to test his strength. And it would be a wise precaution because he and Jones barely spoke to each other now. He said to Wilsher, 'There may have been some purpose when I was away from Washington but with the book virtually completed I don't see a lot of point in Harry Jones staying on there.'

Wilsher looked steadily at Hawkins for a few seconds. 'Then we'll bring him back.'

'Not been particularly impressed by anything he's done there anyway,' added Doondale and Hawkins realised he was going to have to be careful not to get carried away with what he now appeared able to do. What would Jones say, with such an opportunity, he wondered. He said, 'Washington's a difficult place to understand, at the beginning.'

When the lunch ended he went with Wilsher to his office and the editor insisted they drink something stronger than mineral water, to celebrate.

'Gather things aren't good between you and Harry,' said Wilsher.

'There've been disagreements,' said Hawkins.

'I've had a lot of telephone calls.'

It was the way Jones would have acted, Hawkins supposed. 'It's unfortunate,' he said.

'In view of the feeling, I thought you were generous upstairs with Doondale: you could have sacrificed him.'

'I'm not interested in sacrificing people,' said Hawkins.

'Do you realise what happened up there?' demanded Wilsher, offering the brandy.

'What?' said Hawkins.

'You've just been given the key of the door,' said the other man. 'It's going to take a hell of an earthquake to cause you any damage.'

He knew Harry Jones was an enemy and someone who had – and probably still would – do everything to undermine him but Hawkins felt no satisfaction when the recall message came from London. Conscious of the hypocrisy but wanting to make the gesture he bought the man a farewell lunch at the Washington press club and afterwards wished he hadn't because the food wasn't good and it was an awkward meal anyway, Hawkins aware of his own insincerity and Jones, recognising the defeat, deciding the attitude was patronising, worsening the encounter.

There had been a personal letter of gratitude for what Hawkins had done – carefully filed, in the manner of his father – but no personal contact between him and the White House since his arrival back from the rescue mission. Still with time before Eleanor's return to the capital and with a planned tour brief enough not to keep him away beyond that time, Hawkins decided to test Harriman's invitation personally to accompany him on the re-election campaign. The President kept his promise. Within a day of Hawkins making his application through Volger, the press secretary came back to say a place was available on Air Force One for the southern states visit to Florida, Tennessee and Georgia.

Hawkins was surprised at the apparent change in the President. The confidence of long political experience had gone; Harriman had the demeanour of a beleaguered man, able enough to glad-hand among party faithfuls but hesitant and uncertain whenever he had to address an audience and even more ill-at-ease when appearing before press or television. During the flight back to the capital Harriman invited him up into the front of the aircraft, into the Presidential quarters, for another exclusive interview. This time there were no disclosures. Harriman tried ineffectually to dismiss Peterson's lead in all the popularity polls and tried, just as unsuccessfully, to fudge a response to Hawkins' questions about possible progress with the disarmament talks in Geneva.

The freed men had been brought directly from California to a psychiatric clinic in Maryland where Hamilton was the principal, and even when he was on the dispirited presidential tour Hawkins maintained weekly contact with the man there, anxious for the opportunity as soon as they recovered sufficiently, to interview the POWs, aware of the gap in the book and determined to fill it. It was during one of those calls, the week he got back to the capital, that Hamilton made the request and Nicole agreed when Hawkins passed the suggestion on to her. He drove her out from Washington and Hamilton thanked them both for coming, the woman particularly, sitting them in his office and explaining the idea behind the recall experiment.

Gauging the doubt in the psychiatrist's voice, Hawkins said, 'Isn't the recovery good?'

'Off-the-record?'

'Of course.'

'There isn't a recovery at all,' disclosed Hamilton. 'Physically they're in great shape now, but mentally there's been no improvement at all, not from the time I first saw them aboard the freighter.'

'Surely that's not possible,' said Hawkins.

'There are stated cases in the textbooks, where people have been subjected to fear or brutality bad enough for their minds just to shut off. And we've been hampered all the time by not knowing what drugs were used, in their indoctrination . . .' Seized by a sudden idea, the psychiatrist turned to Nicole and said urgently, 'Did you ever see anything administered during their sessions?'

Nicole shook her head, deflating the sudden hope. 'I took them to the place where they were held: sometimes there were men who appeared to be doctors . . . equipment. But I never saw any actual treatment.'

'No bottles . . . phials with a name you can recall?'

'Yes,' she said. 'Both . . .' Seeing the expectation in the man's face she deflated again. 'But I can't remember what was written on them . . . what they were . . .'

Hamilton made a helpless gesture, as if she'd confirmed something. Although they had gone minutely through it, Hamilton said, 'Sure you understand what I want you to do?'

'I think so,' she said.

'Because they only respond to Vietnamese that's obviously how the sessions have been attempted but all we've got is obedience to the simplest of orders. This is extreme: desperate if you like. But it just might get through.'

'I'll do anything to help,' promised the woman.

'I've sent them on ahead because of the travelling problem,' said the psychiatrist. 'We'll go by helicopter.'

By air it took only thirty minutes to get to Camp Peary. They went through security checks Hawkins remembered from the training sessions in which Nicole had also been involved and then travelled along the familiar side roads to where the camp mock-up still stood. Its permanence surprised him and he wondered if it had been re-erected, just for this. There was an ambulance near the false gate and as they approached the doors opened and orderlies ushered from it Forest, Bartel and McCloud. Hawkins' first impression was of their physical improvement. There were no longer sores on their faces or hands and the sharp-boned gauntness of near-malnutrition was gone. McCloud, in fact, looked positively fat: Hawkins supposed it was difficult to engage them in any sort of fitness exercises.

Keeping to the Maryland clinic rehearsal, it was one of the orderlies who gave the Vietnamese instruction for the men to move. Dutifully they formed up and entered the camp, heading towards a prefabricated building Hawkins couldn't remember from the previous occasion. There were already Christmas toys in the windows of Washington stores, and the stiff, mechanical movement reminded Hawkins of a display tableau.

'Is this how it was?' Hamilton asked Nicole, gesturing to the construction. 'It was built on the recollections of the incursion squad who went in.'

Nicole frowned at it, then nodded. 'As I remember,' she said.

Inside the building there was a slightly raised dais with a table and two chairs and other chairs set out in front. The Vietnamese-speaking orderly told them to sit but Nicole said at once, 'No. Always they stood. To attention.'

At the gesture from Hamilton the order was corrected and Forest, Bartel and McCloud obediently stood, hands by their sides. Hamilton escorted Nicole to the table and sat beside her, to prompt. There were two tape recorders and microphones already arranged.

'It's got to be very quick,' said Hamilton, reiterating a Maryland warning. 'I dare not risk a prolonged interrogation.'

'I understand,' assured Nicole.

'Ask them if they know who you are,' he instructed quietly, starting the machines.

She did. There was no response.

'Again,' said the psychiatrist. She repeated the question, voice more demanding. The three stood mute before the table.

'Speak directly to Colonel Forest. Ask him if he remembers giving you the dog-tags?'

This time Nicole asked the question three times but again there was no reply.

The building was only a shell, an attempt to create an effect for their hoped-for recovery and without any heating, despite which Hawkins found himself sweating. Hamilton was, too.

'Ask them if they know where they are?'

Nothing.

'Ask them their names.'

Nothing.

'Ask them their names, rank and serial number.'

Nothing.

'Tell them that unless they reply properly to questions, they will be punished.'

Nothing.

'Tell them we are considering freeing them from imprisonment: sending them back to America, to their wives and families.'

Nothing.

'Tell them that unless they reply, their punishment will be to be kept here and not allowed to return.'

Nothing.

Hamilton looked up despairingly from his clipboard, in

Hawkins' direction. He quickly wiped his face with a handkerchief and said to Nicole: 'Tell them they are heroes.'

Nothing.

'That transport is on its way to the camp, to take them away.'

Nothing.

'Tell them that they no longer have anything to fear: they're safe. And always will be.'

Nothing.

Abruptly Hamilton snapped the machine off, humping unmoving over the table in a moment of depression, before looking up at Hawkins. 'It hasn't worked,' he said. He got up just as jerkily as he moved to stop the fruitless recording and came across to where Hawkins stood. 'Do you realise what I've just done!' he said. 'I've just broken every psychiatric rule in the book; tried putting those guys under as much pressure as they were in the proper camp. Risked them to breaking point. And for nothing; all for absolutely nothing.'

'What now?' asked Hawkins.

Hamilton gazed at him steadily and said, 'There isn't anything else: this was the last shot.'

Hamilton's depression affected all of them. It was impossible anyway to talk properly in the returning helicopter, their parting at the clinic was perfunctory and for some way during the drive back to Washington Nicole remained silent. At last she said, 'I wanted so much to help.'

'It wasn't your fault it didn't work.'

'Poor men,' she said, softly. Then, even softer 'Poor wives.'

Caught by her sudden reflectiveness, Hawkins thought there were other improvements that could be made to the book, beyond an explanation of the Chau Phu ambush. Despite the long hours they had spent together, he still didn't feel he had adequately covered the relationship between his father and Nicole: that there was a similarity of gaps, in fact.

'Tell me what it was like, when you were in Saigon?' he said.

'I've told you before. Many times.'

'After Chau Phu,' prodded Hawkins.

'There was little time after Chau Phu,' said the woman. She spoke looking away from him, out at the autumn-yellowed countryside.

'Explain to me again what he said, about getting you out to the West?'

She shrugged, someone weary of giving account. 'I said I would not leave, if my mother would not leave. He said he would wait, until I contacted him that I was ready. And then he would arrange it: that he had important friends and could fix things as difficult as this.'

'Then why the hell didn't he!' demanded Hawkins, as depressed as everyone else by the failure of the recall experiment and further exasperated by another familiar, unscalable wall.

Nicole looked back, into the car. 'But I've told you he did!' she said, equally irritated.

'How do you *know*?'

'That was part of the evidence produced against me, my association with him. Why I was sent to the camps . . . why, finally, I was put with people who'd been with him on the actual mission . . .'

Hawkins looked sharply at her, then coasted the car off the road on to the hard shoulder so that he could concentrate absolutely. 'How do you know that was *why* you were put with Forest and Bartel and McCloud: you've never said that before.'

She looked curiously at him, unable to understand his curiosity. 'They twisted things, during the hearing against me,' she said. 'They called the rescue flight baby snatching; said they had been taking babies away when they should have been left to absorb the new culture. And when there was the last transfer, to Can Tho, I appeared before a committee who refused my release on new evidence and said my re-education would be completed among people who had actually been upon that mission . . .'

'Wait a minute,' stopped Hawkins. 'I don't understand that.'

'At that last hearing,' said Nicole patiently, 'they said I was being sent to Can Tho as an additional punishment because since the time of the re-unification of Vietnam people in the West . . . important people . . . had tried to get me out. Made applications.'

'Which people?' demanded Hawkins.

'Your *father*,' she said. 'Although I hadn't told him I wanted to come out, which was the arrangement we decided upon, he had made approaches.'

'They *told* you this!'

'It was the *evidence* that got me imprisoned for a further year, as an additional punishment.'

'They could have fabricated it,' disputed Hawkins, 'you said a moment ago they twisted things.'

She shook her head, sad at his misunderstanding. 'Part of the evidence referred to my not leaving, because of my mother. Only he and I knew that: so only from him could they have learned it.' She reached across, feeling for his hand in a comforting gesture. 'That's why I know your father wanted me with him. Loved me. Just as I loved him. He was such a wonderful man: needed me so much . . .'

Hawkins felt out, starting the car and regaining the road. *My father was a hero, a man of whom I'm proud*, he thought.

It remained a good beginning for the book.

Chapter Twenty-Eight

Eleanor telephoned the day she got back to Washington and it was her suggestion that they meet that evening. She picked him up in her car, as she had before, at Columbus Circle and this time she was smiling, at some secret knowledge, not irritated at the subterfuge. Wanting her – just to touch her – after so long, Hawkins tried to lean across the car. She pushed away, but not offended, and said, 'Wait!'

'I can't.'

'You've got to.'

'Why?'

'Because I say so. Tell me what's been happening?'

'I've missed you like I thought it was impossible to miss anyone.'

Eleanor relaxed the touching rule, reaching across for his hand. 'It can't have been as much as I've missed you,' she said. 'God, how I hate all these rallies and dinners and lunches and stupid people! You know how I kept them out?'

'How?'

'By shutting myself off and just thinking of us . . .' She laughed. 'You'll see.'

'Tell me now.'

'No: you haven't told me what's been happening.'

Hurriedly, like a child being promised a reward for good effort, he told her about the London visit and the intended memorial dedication to his father – but not about the relationship established between himself and the proprietor, because it might have sounded conceited – of his brief accompaniment of Harriman and lastly, because it fitted the chronology, of the failed experiment with the freed Americans. The account of the POWs at Camp Peary flattened her happiness.

He was looking across the car and saw her screw her eyes up and she said, 'Jesus, those poor bastards!'

'Yes,' he agreed. 'Poor bastards.' Unhappy at the collapse of her attitude he tried to put buoyancy into his voice as Eleanor drove across the Memorial Bridge and turned south along the parkway, past the Pentagon. 'Where the hell are you taking me!'

'Secret,' said the woman, trying to respond to his effort. 'What did you think of Harriman?'

'Running scared,' said Hawkins.

'That's John's assessment,' she said. 'John thinks he's going to make it: everybody does. Not just John's people. Commentators, too: some not on our side.'

'How do you feel about that?'

'Frightened,' she said.

Breaking a self-imposed rule and straying outside the box Hawkins said, 'What would that mean to us?'

'I don't want to talk about it,' said Eleanor, abruptly, slamming the lid back into place.

They drove by National Airport and Hawkins, trying further to lighten the heaviness, said, 'You trying to run away with me?'

'What a great idea that would be,' she said. 'How's the book?'

'Almost ready,' said Hawkins. 'It was easier before, as a simple biography of my father. Somehow starting with the Vietnam rescue, which is what everyone now wants, seems to make it uneven. And I wanted very much to interview the prisoners; so much hinged on that . . .' He looked across the car at her. 'I made a fool of myself with John,' he admitted. 'Accused him of something he didn't do.'

'Makes a change for people to think he does things he doesn't,' she said. 'Christ, he's a cynical bastard.'

'He certainly used the release and the homecoming,' said Hawkins. 'I looked for you, in California.'

'I wouldn't go,' disclosed the woman. 'He asked me to: said it would look good, with me up there beside the weeping wives and I told him to go to hell. He wanted to take John Jnr and I said no to that too.' She hesitated, returning his look. 'I knew I'd see you there and we wouldn't be able to be together, which was another reason for not going.'

They entered Alexandria and at the first available turn made a left towards the river. Almost at once Eleanor turned left again, to run parallel between it and the main road and pulled into a parking spot. 'We're here!' she announced, triumphantly.

'Where?'

'We can't use your house, right?'

'Right,' agreed Hawkins.

'Or an hotel, for obvious reasons?'

'Right.'

'So I've taken an apartment.'

'You've done what!'

She giggled, pleased at his surprise. 'Actually it's a divided house, but we've got separate entrances.' She handed a set of keys across the car. 'Just us,' she said. 'No one else.'

The entry was at the side of the house. Stairs went immediately upward, directly inside the door and when he put the light on Hawkins saw flowers arranged on two small tables and bottles on a tray, through the break into an open-plan kitchen.

'I came here this morning,' she said. 'I wanted everything to be just right.'

Hawkins turned to her, feeling out, and said, 'Do you realise it's been weeks since we've even kissed?'

'Seven weeks, four days and two hours,' she said. 'I've counted.'

They came together urgently, clumsy in their anxiety, fighting at each other, stumbling and then wanting to go down. Her pants tore but not completely and so Eleanor ripped them further, as impatient as he was, dragging him into her and then bursting before he did, thrusting up at him in the demand that he match her and then when he didn't bucking to make it again herself so that she could match him, which she did. They undressed each other where they lay, clumsy still, eating and feeling and exploring and made love again and this time he established the demands until Eleanor said 'I'm getting bruised against this damned floor,' and got up and led him, breasts bouncing, into the bedroom. When he protested he couldn't do it again, so soon, she pulled his hand into her and then cupped and played with him, head low, and when they made love for a third time he groaned at the soreness and said, 'Christ, are you trying to make up in one night!' and she said 'Yes,' head muffled again.

'I don't care what happens,' said Hawkins, quite abandoned. 'I just want you.'

'I want you too,' said the woman.

It had not been necessary for the Patton Corporation to go through the formalities of formal government tender for a long time, several years in fact, but the invitations were suddenly no longer automatic, arriving instead in an unaccustomed bureaucratic manner and just as bureaucratically – and unaccustomedly – refused. Four ships and seven aircraft obscurely owned by the Corporation had been exclusively but even more obscurely built for and utilised by the Central Intelligence Agency and within a month of the Vietnam rescue – and revelation by Peterson – their utilisation began to diminish and when Patton tried to establish contact at Langley with people who earlier responded to every telephone approach he was told they were unavailable. His calls were never returned. Neither were they answered from the Pentagon, to which he had been summoned from the aircraft carrier in the South China Sea and where he had repeated his regret at the error of contacting Peterson, an aberration for which he could offer no logical explanation, beyond that he had made on the boat to Hawkins and Jordan and which seemed emptier in the hollow, sterile rooms of the military headquarters than it had at sea.

Patton was not a business fool, just an otherwise preoccupied man. He recognised easily enough the cause – although he did not know the cure – of the sudden, crippling closure of Washington against him. And he didn't bother himself about the cure, involved completely with Sharon.

To be nearer the husband she had so recently – finally – accepted as dead Sharon Bartel moved to Washington, all the doubt and guilt returning in contrition, refusing Patton's offer of accommodation at the Four Seasons or the Jefferson or the Mayflower and staying instead at the nearest Howard Johnson motel to the Maryland clinic – further contrition – at a reduced rate for longterm occupancy.

Like other people around Washington he attempted to contact during the working day, initially Sharon refused to take his calls but then the stupidity – and the need – overcame her and she agreed to their meeting. Patton was cautious, as he had always been cautious in their relationship, letting her make the pace, something else he'd always done. He knew she wouldn't sleep with him, not now at least, so he never made the suggestion, not wanting to in the circumstances either. He kept a permanent suite at the Four Seasons and made it clear it was hers whenever she wanted it and spent every weekend there and commuted to it at least twice a week and had a chauffeur collect her and return her – unaccompanied – to the Maryland motel after every visit. It was a papier maché existence and Patton endured it because he loved her and hoped, because she accepted his visits and consented to be with him, that she loved him too, as false and unreal and as papier maché as the situation was.

She'd used her own key to the Four Seasons when he arrived and it was obvious she had been there for more than one day and Patton, always a man of omens, was more cautious than he'd ever been. She seemed grateful at his suggestion that they eat in the suite, although when the food arrived she only picked at what she'd ordered, disinterested, and he played with the food as well, more disinterested than she was.

'They say it's irreversible,' she said. 'That all the tests and experiments have been done and he'll never get better.'

'We'll get different doctors,' promised Patton. 'They can't say that.'

'Why?' she said.

'Why what?'

'Why are you being like you are: being so wonderful?'

'He's your husband.'

'I still don't understand it.'

'I love you,' said Patton.

'That makes even less sense.'

'We'll get every advice, every expert,' said Patton. 'If they think anything – anything at all – can be done, then we'll see it's done. If, after all the tests, every examination, the verdict is that it's truly irreversible then I still want to marry you.'

'Could I do that?' she demanded. 'Could I abandon someone I once loved, someone I don't even now know!'

'That's for you to decide,' said Patton.

'That's the problem,' said Sharon. 'Deciding.'

Chapter Twenty-Nine

The period leading up to the final election was for Hawkins one of several levels, with the election itself intruding like an echo, through each of them, a nagging reminder that he did not want to acknowledge. The most settled – but contrarily the most endangered – level was with Eleanor. The Alexandria apartment allowed a settled pattern to their relationship, a normalcy for something that could never be normal. In the final months Peterson travelled constantly, rarely with her and so she and Hawkins were able to be together practically whenever they wanted. And Hawkins wanted them to be together as often as possible.

Alexandria ceased being a place just for lovemaking. It became for Hawkins more of a home than Maryland Avenue, in which Nicole and Ninh established themselves as housekeeping custodians – another level – so successfully that he finally dismissed the housekeeper his father had engaged and formalised the roles they had taken upon themselves.

He took the manuscript to the Alexandria apartment for the final polishing and eventually consented – later recognising that he'd brought it there in the first place for her approval – to Eleanor reading it. Which increased the impression – a game they were both eager to play, not wanting it to end – of normal domesticity. When the winter chill came he would set the fire and Eleanor sat at his feet, marking the biography page by page and when they talked it through there was rarely an occasion when he didn't feel her criticism had made an improvement.

'It's good,' she said, on the October day that she finished. 'Very good.'

'There's the gap,' said Hawkins, objectively. 'Always the gap.'

'You've covered it,' she said, knowing what he meant. 'You've described the film; the intensity of the attack. John and your father – all the survivors – saw them fall and wrongly thought they were dead. Everyone's conceded the mistake; and accepted it. I find it easy to understand.'

'It still worries me,' said Hawkins.

'Sure that's what's worrying you?'

'What else?'

'It's time to let it go, darling: to let somebody else see it and make an independent judgment,' she said.

She'd come to know him very well, Hawkins conceded. 'Maybe soon,' he said.

She shook her head. 'It's finished,' she insisted. 'There's nothing more you can do to it. Don't be frightened: it's a wonderful book.'

'I guess you're right,' he said.

'John called from Chicago last night: said he thinks I should spend more time with him now that the election is so close.'

'Rampallie called me, too,' said Hawkins. 'Asked when I was going to catch up with them. I've been putting it off but I suppose I've got to go.'

'We've both been putting things off,' she said.

'So let's stop,' he said, suddenly urgent. 'Let's talk about it. I want to marry you. I want you to divorce John and stop all this buggering about and for us to get married.'

'Don't!' she said.

'Don't you want to?'

'You know the answer to that.'

'Then what's the problem.'

'You know the answer to that, too.'

'OK, so for a time it would be difficult. But it wouldn't last for ever.'

'Yes it would, darling. And you know it,' she said. 'We wouldn't be able to move without having cameras and microphones and reporters all around us; you of anybody should know that. It would be a permanent goldfish bowl. Do you think you could stand that?'

'Yes, don't you?'

'I'm not sure: I'm really not sure that I'm that strong.'

'So what's the answer?'
'There isn't one,' she said. 'There never has been.'
'There's got to be!'
'Not now,' she pleaded. 'Let's talk about it later.'
'Which is putting it off again.'
'I said I'd go up to Chicago on Saturday,' she announced. 'There's the last tour, of twelve States. He wants me along all the time.'
'And I'll be there, watching.'
'Don't you think I haven't thought of that,' she said angrily. 'Don't you think I'll be watching too, seeing you and being polite to you and aching for you and not being able to do anything about it! Don't forget that's why I didn't go to the California homecoming.'
'So this is over?' said Hawkins, gesturing around the apartment.
'For a while,' she said, in flat resignation.
'Only for a while?'
'Stop trying to back me into corners,' she protested.
'I want to back you into a corner: in a corner you can't run.'
'Don't spoil tonight,' she said. 'Let's shut the world out tonight.'
But it was spoiled. There was a barrier between them and each tried too hard to push it aside and their lovemaking was clumsy and unsuccessful. The habit had developed of their remaining throughout the night but she suddenly got out of bed and said she was going back into Washington and Hawkins put up only a token argument because he didn't want to stay either. When it came time to leave neither appeared to know what to say to the other and so they said nothing, kissing cursorily and hurrying to their separate cars.
Hawkins joined the Peterson campaign two days after Eleanor, in Milwaukee. The senator extended every facility, insisting upon Hawkins personally accompanying him in aircraft and buses and even limousines, in two of which he was actually close to Eleanor, in adjoining seats. Throughout they maintained the charade of acquaintances and Hawkins managed to sustain the performance for a week, leaving the cavalcade in San Diego, grateful to be away but missing her by the time he was fastening his safety belt.
Recognising it as a cliché for such a situation, but one that

might help anyway, Hawkins sought to lose himself in work. He accepted an invitation from Volger to accompany the President on one of his concluding tours, deciding with growing sadness – because he'd admired the man as a consummate politician but more because he knew his personal situation would be made easier if he gained a second term – that Harriman's lacklustre, unsure husting performance was worse than on the previous occasion he'd travelled with the man. The tour over, Hawkins committed himself to the final, decisive editing of the manuscript and at last, with reluctant nervousness, airfreighted it to London. After two weeks there were enthusiastic messages of congratulation from the editor and the proprietor – coupled with an instruction to return for the unveiling of the sculpture, for which a date had been fixed – and two weeks after that there was an equally enthusiastic response from the English publisher.

Eleanor had been right in guessing his apprehension at the professional reaction to the book and Hawkins was relieved at the praise. But it was a passing satisfaction, because the cliché hadn't worked.

It was obviously too risky for Hawkins to attempt to contact Eleanor in the protected depths of the Peterson camp but he expected her to find a way of calling him. But she didn't. And as the final weeks passed the depression made him unreasonably irritable with everything and everyone, particularly Nicole and Ninh because they were always in the house. Finally the woman openly asked what they had done to offend him and whether he wanted them to find somewhere else to live. He was tempted to use the opportunity because the arrangement was becoming too settled and he wanted them to stop depending upon him so completely – and for himself to depend upon them, just as completely – and find some other accommodation but he was embarrassed at causing them the misunderstanding, so instead said they had done nothing to offend him, which was true, and that he didn't want them to find somewhere else, which wasn't. He remained irritable and they remained unconvinced and subdued.

The election provided the excuse for him to at least be near her again. He joined the campaign for the last days and was received with the same special attention as before.

Hawkins was surprised, in such a comparatively short time,

how things had changed with Peterson. Before the senator had been a contender, the Party choice with a chance but still someone who had to unseat an incumbent. Now, with the campaign at an end and the polls practically unanimous and the soundings being analysed from every State there was a discernible confidence, the atmosphere and behaviour around the man already presidential.

The night after his arrival Hawkins was invited to a private dinner, still a large affair, for the party hierarchy and the campaign workers most closely involved, like Rampallie and Elliston. At the reception before he was aware of Eleanor's indication and saw her mingle cleverly in his direction and he matched the subtlety, moving from group to group until finally they were alone, at one side of the crowded room.

'I've missed you,' he said at once, wishing there were words better to describe his longing.

She looked hurriedly around, to ensure they could not be overheard, and said 'Christ, I'm exhausted.'

'I said I've missed you.'

'Not here,' she said. 'It's too dangerous.'

'You didn't call.'

She frowned at him. 'You've seen what it's like! How the hell do you think I could have called. I can hardly go to the bathroom by myself.'

'When?' he demanded.

'I don't know!' she said.

'Promise you'll call?'

'If I can.'

He caught Peterson's eye on the far side of the room. The senator smiled and waved and Hawkins waved back.

'This is awful,' said Eleanor. 'I don't think I can handle it.'

'There's nothing to handle at the moment,' said Hawkins.

'John's coming across,' she warned.

Hawkins turned to greet the man. He decided that Peterson had changed, like the ambiance around him; he seemed almost physically bigger and the gestures and expressions were expansive, as if he were performing for an audience.

'Expected to see far more of you,' said Peterson.

'I intended to spend more time,' said Hawkins, uncomfortably. Snatching for an excuse, he said, 'Got caught up with the book far more than I expected.'

'How's that doing?' said Peterson, with a politician's ability always to involve himself with the person to whom he's talking.

Hawkins was conscious of Eleanor moving uncomfortably beside them, seeking an escape from the situation. He said 'It's finished: everyone seems pleased with it.'

'I'd like to read it,' said Peterson.

'I'll make sure you get a copy,' promised Hawkins, imagining nothing more than politeness.

'Why don't I do a Foreword for it?' the senator suddenly offered.

'What?' said Hawkins, too aware of Eleanor's closeness properly to concentrate.

'I knew the man,' said Peterson. 'Why don't I write some sort of introduction . . .?' He smiled, with apparent shyness and said, 'That's if you'd like me to, of course.'

'Naturally I would,' said Hawkins, confused by the offer. 'That would be very good of you.'

'Fix it through Rampallie,' said Peterson, gesturing vaguely in the direction of the room. 'Get a copy or proofs or something to him.' He grinned down at his wife. 'Time to mingle with the faithful, darling,' he urged. To Hawkins he said, 'Good to have you with us, Ray. Going to be an exciting few days.'

Hawkins didn't find them exciting. He found them a strain, sometimes almost an agony, close but yet always separated from Eleanor, constantly favoured and befriended by her husband, which made everything worse. There was only one further occasion when they managed a private conversation, a snatched whispered affair on the plane taking them to New York when she promised to telephone, whenever she could. There was an entire floor of the Plaza reserved for the night of the election and with his special status Hawkins was welcomed into the suite where Peterson was sitting, monitoring the returns state by state from a battery of television sets. There were other selected correspondents and a television crew and two still photographers and for their benefit Eleanor sat beside her husband, holding his hand, and it was at her that Hawkins was looking when the decisive states of Florida and California declared, giving Peterson the Presidency. Briefly – only seconds because for the benefit of the cameras Peterson swept her into an embrace – Eleanor turned back, answering Hawkins' look: her face was quite without expression.

It was a long time before Hawkins was able to make his way through the crush but finally he did.

'Congratulations,' he said. Quickly he added, 'Mr President,' and Peterson laughed and said, with perfect modesty, 'It's going to take some getting used to.'

'Congratulations,' Hawkins repeated, to Eleanor.

'Thank you,' she said. There was no expression in her voice, either.

Eleanor telephoned a week later, at once halting any hope of their meeting by saying that after the strain of the campaign – and before the inauguration – Peterson had decided upon a family holiday in the Caribbean, which John Jnr had enjoyed last time, and which meant they were going to be away from the capital even longer than she had expected.

Hawkins tried work again, with a written list of publishers' queries to answer before the London editing, which he decided to do during the visit for the unveiling. He was never able to recall the moment the idea consciously came to him. Suddenly, abruptly, he was thinking about it, as if it had been a consideration for a long time, which it hadn't and because it hadn't he became annoyed with himself because it was so obvious and therefore it should have occurred to him as a journalist, irrespective of any work upon the book. His mind echoed around one word – physically – like the ringing sound of the telephone when he called Hamilton at the Maryland clinic.

Hesitantly – an amateur approaching a professional, even though he'd been involved in the other experiment – Hawkins offered his suggestion to the other man, anticipating annoyance at the impertinence.

'I didn't know a film existed,' said Hamilton.

'It didn't occur to me to tell you, because I thought you would know,' said Hawkins. 'I'm sorry.'

'If Camp Peary didn't work, then I'm not sure about this,' said the psychiatrist. 'Sharon Bartel had her husband away from here for over a fortnight, trying other psychiatrists, other tests. Nothing succeeded: they're becoming guinea pigs.'

'It's a chance,' persisted Hawkins. 'Shouldn't it be tried, if it's a chance!'

'Yes,' conceded Hamilton. 'If it's a chance I suppose it should be tried. Everything else has, by everyone.'

'There's nothing that can be done,' said Patton, exasperated. 'Nothing!'

'I know that,' said Sharon Bartel.

'Then what's the point?'

'There isn't one,' admitted the woman. 'I just can't.'

She sat nervously on the edge of the couch in the Four Seasons apartment. Patton stopped pacing the room, actually kneeling before her so that she had to look at him. 'Your husband will never be able to leave the hospital,' he repeated, patiently. 'Never be able to recognise you: know you. You're achieving nothing, by staying married to him. Not helping him. Nothing. There's no purpose in martyrdom.'

'I'm not trying to make myself a martyr.'

'What then?'

'I can't put it in words you'll understand,' said Sharon. 'I just know that stupid though it may be I can't abandon him.'

'You wouldn't be abandoning him,' argued Patton. 'He doesn't *know*.'

'I'd know.'

'So what are you going to do!'

'Stop seeing you,' she announced simply. She didn't try to avoid his eyes, staring straight at him.

'What!' he said.

'Stop seeing you,' she repeated. 'I'm not strong and if we went on as we were before I know I'd weaken and finally agree and I don't intend to do that.'

'You don't realise what you're saying!' said Patton. 'You can't do!'

Sharon started crying at last, tears edging down her face and she sniffed angrily, trying to stop the collapse. 'Oh I do, my darling,' she said. 'I realise just what I'm saying . . .' She fell forward, against Patton's shoulder. 'Why!' she said, her voice muffled. 'Why did he have to come back, after all this time?'

Chapter Thirty

CBS were as helpful as before but the cooperation took longer this time because they would not let the master-tape out of their possession and instead made a copy of Lind's film of the Chau Phu episode, so it was a week after Hamilton's reluctant agreement that Hawkins drove out again to the Maryland clinic. It was one of the first real days of winter, the wind bitterly cold and snow threatening from the leaden clouds, like the weather had been when Eleanor first came out to Maryland Avenue, with a red nose and the woollen hat he always pictured her wearing pulled low over her ears. Hawkins thought how easy everything had been then, compared to how difficult it was now. How difficult? he demanded of himself, in hopeful desperation. Several Presidents had conducted later recorded affairs in the White House; their wives too. Not impossible then. Just difficult. But was that all either of them wanted, an affair? He certainly wanted more. And he wasn't sure if there were a precedent for a President's wife divorcing her husband during the period of his incumbency to marry somebody else. Eleanor had been right, during their parting argument in Alexandria: if she agreed they'd be hounded and pursued when it became public, no matter where or how they attempted to hide. He didn't give a damn, Hawkins determined, still clutching at the desperation: if Eleanor agreed he was prepared to make whatever sacrifices it took.

Hamilton's reluctance was obvious, when Hawkins got to the clinic. 'They *are* becoming guinea pigs,' he said to Hawkins. 'It's hardly ethical, to go on like this.'

'It's surely not as traumatic as Camp Peary?'

'I regret doing that now.'

'It might have worked.'

'It didn't: neither did anything else that was attempted separately upon Bartel.'

'How are they, physically?'

'Fine,' said Hamilton. 'That's part of the tragedy: there's every chance of their living for years, if you can describe their existence as living.'

'No trouble from their wounds?'

Hamilton shook his head. 'Bartel had that severe cut to his face, but that's healed completely. So's the cut Page got when they were escaping.'

Hawkins was about to continue the questioning but changed his mind, considering the film experiment the first priority: if that worked, the other questioning would be easy.

Hamilton insisted on seeing the recording first, just with Hawkins, and at the end shook his head doubtfully and said, 'Christ, I don't know: I really don't know. That's as dramatic as hell.'

'Which is why it might work,' pressed Hawkins.

A long silence stretched between them, which Hawkins didn't try to break because he knew there wasn't any further persuasion he could offer. Finally Hamilton said, 'OK: we'll give it a shot,' and Hawkins felt the relief move through him.

Hamilton gave the instructions into an intercom and Hawkins helped him move the video and the television into a larger area, a lecture annexe off Hamilton's main office. They were still assembling the system when Forest, Bartel and McCloud entered. There was a uniformed attendant with each man: one of them was Asian-featured and Hawkins guessed he was Vietnamese. The POWs were unchanged since the last time he had seen them, at Camp Peary, except for McCloud, who looked even heavier than he had then. They sat, to the Vietnamese instruction, gazing blank-eyed at the unlit television screen, hands cupped identically in their laps.

'Page wasn't involved at Chau Phu,' said Hamilton. 'There didn't seem any point in including him.'

There would have been a point, for Hawkins' purpose but not for that which Hamilton was trying to achieve. Hawkins decided there was nothing to be gained by arguing.

'Guess we might as well start,' said Hamilton, moving to the video set alongside the television screen. As he depressed the button, the Vietnamese attendant drew the curtains, half darkening the room.

There was a momentary flicker, a technical jump of numbers from five downwards as the film was counted off and then an image formed on the screen, the picture of John Vine giving the editing instructions in front of the assembling group. Hawkins only half looked at the film, wanting merely to know the sequence, concentrating more fully upon the freed prisoners.

'This is Vietnam, in the final, agonising days of a long and bloody war . . .' Hawkins heard the dead reporter say. Forest, Bartel and McCloud sat square-shouldered, light from the television screen reflecting whitely on their faces. An occasional blinking was the only mobility in their faces.

The interview with the man now the President of the United States came into focus. When Peterson talked of the abandonment of Vietnam the Asian orderly said 'Right on' and Hamilton said 'Shut up!' annoyed at the interference.

When a reaction came, Hawkins missed it. He'd turned briefly away, fully towards the screen, when there was a stir, something like a foot movement. Hawkins jerked back, not knowing if it were Bartel or Forest but was sure it was one of them. Hamilton was staring too. Suddenly there was a more positive reaction, definitely from Bartel, a deepthroated groan as the picture showed the storm of dust and debris when the helicopter landed and the Green Beret group leapt out.

'A helicopter again, just like on the freighter,' whispered Hamilton, excited, all reluctance gone now.

The groaning continued – unlike any sound Hawkins had ever heard a human voice make before – at the entry into the abandoned, bare orphanage and Forest responded this time, in another unintelligible noise but loud, like a warning, at the shot of the line forming to ferry the children into the helicopter.

Forest positively shouted again, louder this time, at the moment of ambush and abruptly, so unexpected that neither orderly was able to prevent it, Forest and Bartel stumbled upwards, as if they were going towards the screen.

Forest's attendant was able to get to him in time, grappling with him, but the man looking after Bartel wasn't quick enough and Bartel collided with the arranged chairs directly in front of where he had been sitting and fell over them. Bartel was awkward and uncoordinated, making no effort to save himself and went heavily against the ground. The breath was gouged out of him, cutting off the groaning as if a switch had been thrown. Hamilton thrust back the curtains, better to see the three. McCloud, who until now had not left his chair or made any noise, slumped forward, breath rasping into him and Forest stood with the attendant's arms around him, still making the gutteral sounds. Bartel, winded, appeared semi-conscious.

'Quickly!' ordered Hamilton. 'Into an examination room.'

The only concern was for the three men, Hawkins ignored. He stood back as the patients were moved, half-carried, towards a linking door into the room with beds. They were dishevelled, their hospital gowns scarcely on them and Hawkins looked and decided he wouldn't have to ask any more questions because he could see for himself. The neglected television screen, the film over, snowed whitely in front of him. Without – initially – positive thought Hawkins went to the machine, rewound the tape and played it through again for himself. When it finished he was more wet-eyed than he had been in the New York viewing theatre, although not because of what he had seen on the screen but because he was sure that he knew.

It was a further hour before Hamilton came back into the room, face furrowed, so deep in thought that he seemed momentarily surprised to find Hawkins still there.

'Bartel's cracked a rib,' said the psychiatrist. 'We've had to strap it.'

'What about the reaction?' said Hawkins.

'It's the first positive thing we've got,' admitted Hamilton. 'I'm not happy with it, because of the trauma it caused. But I think I've got to try it again. Can I keep the tape?'

'For as long as you like,' said Hawkins.

'There's a connection with the reaction to the helicopter, aboard the freighter,' said Hamilton, more a reflection than a continuation of the conversation. 'I think, blocked though they are, they regard the helicopter as something that might have prevented their rescue.'

Now that he knew the truth, how was he going to prove it? Hawkins asked himself, driving back to Washington. He was going to have to, some way: his future with Eleanor depended upon it.

Hawkins knew they misunderstood, imagining it a continuation of their previous misunderstanding about his irritability over Eleanor, but he didn't attempt to correct it, seeing the way as them – or rather Nicole – being as unsettled as possible. Ninh, nervous, remained in the kitchen in which Hawkins demanded he stay and Nicole, just as apprehensive, hesitantly entered the study, looking enquiringly at him for direction. He indicated the couch in front of the fire, positively closing the door behind her and sat forward in a bordering chair, so that he could look directly at her.

'You lied to me!' he said, immediately accusing. He winced, inwardly, at the bullying.

'No!' she said, as immediately defensive.

'Tell me again,' said Hawkins. 'Tell me about my father.'

'I've told you. A hundred times I've told you!'

'Again,' he insisted.

'We met at the Cercle,' she started, uncertainly. 'He was not normally a man to whom I would have been attracted. But there seemed about him an attitude of gentleness: there wasn't much gentleness, in Vietnam . . .'

It wasn't what he wanted but Hawkins decided to let her continue, until she reached the point.

'. . . He asked me out and at first I refused and then he asked me again, so I went, just curious, that's all. It was a restaurant on the Rue Catinat: he was very proper which was something else which was unusual . . .' She shuddered, at a memory. 'Men were rarely proper to women, in Vietnam . . . there were more meetings: lunches and swimming parties, at the club mostly. He didn't kiss me for a long time: attempt to, even. In the end it was I who kissed him . . .'

Hawkins was burning with discomfort at what he was forcing the woman to do; it was like looking through the keyhole into somebody else's bedroom.

'. . . Everything was always in so much of a hurry in Saigon,' she said, gathered up now in the reminiscence. 'War, I suppose, makes everything fast. But he wasn't fast: it was part of a gentleness, a slowness of someone who always had time to care . . .'

Nicole looked directly at him, flushed, and from the redness of her face Hawkins knew she was as discomfited as he was. 'We never made love . . . even slept together . . . until I agreed to move into the villa with him . . .' There was a smile, at the grateful return into memories. 'It was like marriage,' she said. 'A proper marriage where a man respects a woman . . . I wanted the baby so much: more than he did, I think. At first at least. Not after it was born: he was very proud then. We had an American perambulator, with big, inflated wheels and on Sundays went for walks, like ordinary people . . .'

Hawkins brought his hand to his face, this time to control the outward wince. Christ, was the truth important enough for this! No, he thought. Then, immediately, yes. But he couldn't let her continue at her pace, exposing herself so nakedly. It meant taking an enormous chance, risking everything he hoped to discover.

'What did he tell you, about Chau Phu?' demanded Hawkins.

Nicole blinked, emerging from the past. 'You know about Chau Phu,' she said.

'I know what he wrote.'

'That's what happened.'

It hadn't worked, thought Hawkins. 'How was he, afterwards?' he tried again.

'Upset, naturally,' said the woman. 'He should never have gone . . .'

'Why not?' demanded Hawkins, seeing the crack and jabbing out to widen it.

'Nothing,' she said, trying to retreat.

'Why not?' repeated Hawkins, insistently. *People always give more if you appear to have more.* Using the familiar axiom of his father's against his father, he said, 'I know, Nicole: I know it all.'

'Are you surprised then!'

'No,' he said, the only response possible: he tensed, willing her to continue.

There was a heavy, hanging moment of silence. And then she did.

'He *was* brave,' she said, going back to her earlier defensiveness. 'By going, he was brave, as brave as he was supposed to be, from his reputation . . .' Her mouth came together, in a tight, bitter line. 'Damn the reputation! That's all he was doing, in the end: living up to a reputation, one that he hadn't wanted anyway and which became a constant challenge: one he didn't want. And couldn't finally, confront. There was no one like your father; no one. All the wars; every one of them, for so long. Until there were too many. Too much noise. Too much shooting. Too much death . . .' She smiled up. 'And I know it was because of me. I know it looks as if he didn't care: that he didn't make any attempt to get me out but I know that was why he was frightened. We talked about it. He decided he'd done too much, been too lucky. And that his luck was going to go suddenly. That he would be killed . . .'

Hawkins realised that he had unlocked the door but he hadn't expected the clutter within: he frowned, trying to understand what she was saying.

'So frightened,' she remembered. 'At night, usually. That was the worst time. He couldn't maintain the control, not when he was asleep. The dreams would begin and he'd start crying, still asleep and then wake up and realise what was happening and cry more because in the end I was the only person he wanted to be brave with and he would be ashamed, at having collapsed . . .' She extended her arm, an encircling gesture, and said, 'He'd lie against my shoulder and say he was sorry, which was stupid because there was nothing for him to be sorry about.'

'What about Chau Phu?' said Hawkins, trying to find a pathway.

'That *showed* his bravery,' she insisted, defiantly. 'No one knew it but me; and they still don't. But that showed his bravery. Peterson wanted him to go, because of the damned reputation, knowing that whatever your father wrote would guarantee him greater coverage, internationally. But he didn't have to go. He could have surrendered the place, without anyone guessing the real reason. But he didn't. He went to prove himself to himself, and to me, trying to impress me, like a young lover . . .'

'Did he?' intruded Hawkins.

She frowned at him. 'You know what happened,' she repeated.

'Tell me what he told you,' said Hawkins.

Hawkins pressed forward, intent for any lie or deviation

from anything he had read or watched on film, or heard from Peterson and Patton. But there was nothing. There were minimal discrepancies, just as there had been minimal discrepancies between what Peterson and Patton had told him during the interviews and their recorded evidence at the earlier enquiry. But like Peterson and Patton they were inconsistencies understandable from the passage of time.

He'd tried to jar Nicole into a disclosure, by the earlier accusation of having lied to him, saying he knew. But she was the one who didn't know. Who hadn't lied, not from the moment of their meeting aboard the rescue ship in the South China Sea.

'He was a good man,' concluded Nicole. 'A wonderful man: you should be proud of him.'

'I know I should,' said Hawkins. So why was he doing what he was? He knew but he refused to allow himself the answer, going down into the familiar basement to compare yet again his father's written account and the interviews with Patton and Peterson against what had been said at the initial and latter enquiry. That was where he was when Eleanor called, wanting to meet him. Urgently – excited – Hawkins agreed upon Columbus Circle again.

Eleanor appeared uncertain of a direction and Hawkins said, 'It's a pity we gave up Alexandria,' and she said, 'No, it wasn't.'

She took the car over the Memorial Bridge but turned north, for the parkway to take her on to the peripheral road around Washington, a drive without an end.

Hawkins said, 'Nothing's changed.'

'Don't be ridiculous,' she said.

'About the way I feel, I mean.'

'It's impossible,' she said. 'Do you know I had to give Secret Servicemen the slip to make this meeting tonight! The Secret Service! They think I'm at home in bed. It's worse than I ever thought it would be: horrible.'

'I meant what I said, about wanting to marry you.'

'That's impossible, too.'

'Why?'

'You know *why*, for God's sake! Because I couldn't stand the pressure and we'd fall out of love and end up hating each other, just like your first wife ended up hating you. And I don't want to hate you. I want to love you.'

'This is becoming one of those silly conversations.'

'It's got to end,' insisted Eleanor. 'That's all it can do. End.'

'No!' he said.

'Have you any idea of the power John's got now: what he could do to us? To you?'

'We wouldn't have to remain in America.'

'What about the children?' she demanded. 'I'd be the erring woman: the whore. He'd keep the children from me. And that would be the least he would do: he'd destroy us, if we tried.'

'There must be a way!' said Hawkins, desperately.

'He's the *President*!' she said, as if she didn't think he understood.

'What if he weren't?' said Hawkins.

She looked quickly at him across the car. 'What sort of question is that?' she said. 'He *is*: there's nothing that can change that.'

'He's not inaugurated yet,' said Hawkins, determinedly.

Chapter Thirty-One

Hawkins knew the pressure and quite cynically used it, wanting more now than the truth he suspected because in the truth he saw the way – recognising it as a desperate, insane way – of keeping Eleanor. But anything was worth keeping Eleanor, no matter how desperate or insane. Patton wouldn't take his call, which didn't surprise Hawkins. He asked that the man return it, accompanying the request with the message that things had changed in undertakings he had given, calculating the threat would frighten the industrialist into responding, which it did, within the hour. The apprehension was loud in Patton's voice and Hawkins used that too, refusing anything further on the telephone, insisting instead upon a personal meeting for which, he said, he was prepared to fly up to New York. Without any choice, Patton agreed.

The same austere, bespectacled woman greeted Hawkins at Patton's Third Avenue headquarters and as he followed her through the openplan work section Hawkins had a passing, casually registered impression that there was not as much activity among the assembled workers as there had been on his first visit. This time Patton didn't emerge from behind his overly large, flag-marked desk. Instead, unsmiling, he remained behind it like a barrier and Hawkins had another impression, that the man appeared smaller, shrunken almost, from their previous interview here and from their time together on the freighter.

The woman waited in the doorway, expecting an instruction to provide drinks, but Patton jerked his head dismissively towards her. The man waited only until the door closed and said to Hawkins, 'OK, so what's the change!'

Always feign more knowledge than you possess, Hawkins remembered. Once more, normally good advice but this was going to be a verbal balancing act without any safety net because Patton was his last chance: his only chance. And because it was his only chance he was going to have to take an even greater one. Hoping Patton wouldn't detect the gesture for what it really was, Hawkins reached into his pocket apparently for a handkerchief, pressing as he did so the start button of the tape recorder, worried that the distance between Patton and himself might be too great for the top-pocket microphone. He said, 'CBS made the film available to me, before I came to see you here initially.'

'You told me that,' said Patton. Curtness didn't suit the man; and sensitive – because he had to be – to nuance, Hawkins detected an artifice in Patton's aggression, a slightly louder-than-necessary tone in his voice. He said, 'They made it available to me again, for a recall experiment at the clinic.'

He stopped purposely at the absolute truth, aware that if the recording worked as he hoped it would – and was to have any value – any subsequent transcript would have to avoid showing him guilty of entrapment.

'What!' said Patton.

'Made it available for a recall experiment,' repeated Hawkins, one foot carefully in front of the other on the high wire. Maintaining the truth and producing the incongruity which had triggered everything for him, Hawkins said, 'I saw them without clothes on, at the clinic: disarranged anyway. Know something I couldn't understand? I couldn't understand how men whom Peterson testified and Blair testified and my father testified and you testified to have fallen under an onslaught of firing didn't have any scars. And they haven't, you know. No one seems to have realised that, but how do you think three men – cut to pieces, you said – don't have any wound marks on their body at all?'

Even that wouldn't have worked if Sharon hadn't rejected him and he'd known she meant it and there was no longer any hope for him in the relationship. Patton felt himself on a tightrope too, in his case one that was fraying at either end. And he saw the tall, angular Englishman as the person about to put his weight on the last remaining strands. The aggressiveness, false to his nature anyway and as forced as Hawkins had guessed it to be, cracked away, eggshell thin. Patton groped up, putting his head into his hands, and said in a voice so muted and so strained that Hawkins was sure the tape wouldn't pick it up, 'Oh God, oh dear God!'

Cautious still, knowing he was some way from safety, Hawkins said, 'Why not tell your side? I want to be fair and know your side.'

It was a long time before there was any positive response, any movement at all; aching, stretching minutes when Hawkins thought he had tripped. Then Patton looked up from between his hands, not held for support any more but splayed, as if he were trying to protect himself from some attack. He said, 'I knew it couldn't remain a secret. I said so then and I said it again when those damned people got out and contacted you and it started all over again. They yelled at me that I was weak and Peterson sneered and said how he was right, when we discovered the mental condition they were in but I knew all along that it would come out, one day . . .'

Hawkins was as torn as the other man, but better controlled. He'd been right and so he'd found a way to win but if Peterson and Blair and Patton had been involved then so had his father, his frightened, nerve-shot father in whose honour he had written a eulogy and to commemorate whose bravery a memorial was shortly to be unveiled by a British statesman. Still balancing, he repeated, 'Your side: tell me your side.'

Patton's head went back into his hands and Hawkins leaned forward, anxious for the volume.

'Awful,' said Patton, setting out awkwardly. 'Everything was so awful: Vietnam was awful. Everyone said it was a half-assed mission, at the morning briefing, brown-nosing to some politician with clout from Washington, out to get his name in the papers . . . didn't think it was going to be so awful, though . . .' The man came up, through the shutter of his fingers. 'I'm not brave,' he said, pleading. 'Never. Can't be. All the time in Vietnam I was so frightened that I didn't know what to do . . . I'd lay awake nights trying to think how to get out of missions but so the other guys wouldn't know I was as frightened as I was . . . managed it, until Chau Phu: Headquarters secondment, ferrying Generals and Colonels around, who sure as hell didn't want their asses shot off because they knew it was all going down anyway and weren't going to become dead heroes. If I hadn't been attached to that goddam Headquarters staff I wouldn't have been picked for Chau Phu anyway. That's how the stupid bastards saw it . . . just another chauffeur job.'

The words were like uneven stones washed along by a suddenly unblocked dam, thought Hawkins. But it would be wrong to try to direct the flow: disjointed as the account was, it was better to let the man talk uninterrupted now.

'Seemed OK at first. I'd been shitting myself – literally wanting to crap – all the way there and it seemed OK. Forest pulled his men into the rescue because of how they found the kids . . . I sent Bartel in because I wanted to get to hell out of there as fast as I could and so I told him to get his ass out of the helicopter and join the rescue line.'

Patton's face came up again, pleading: Hawkins saw the man had started crying, wet-eyed and his nose was running and he wasn't bothering to wipe either. 'Do you understand what I'm saying?' he demanded. 'I'm saying I sent a guy, a friend, someone I knew, in to be captured and then I stole his wife . . .' a hand came out, a stopping gesture '. . . no, that isn't right: not like it was. There wasn't anything between Sharon and me, not before: I didn't even know her. I looked her up from remorse, in the beginning. Wanted to help. Money; stuff like that . . .' He looked at the picture frame still prominent on his desk. 'Didn't fall in love with her for a long time,' he said, very quietly.

Patton fell silent, still gazing at the photograph, and when he showed no indication of continuing Hawkins felt out delicately, to prompt. 'What happened at Chau Phu?' he said.

'Awful, like I said,' resumed Patton, mind blocked with a single word description. 'Felt I was protected inside the helicopter . . . metal around me. Didn't hear the first shots, not as firing anyway, not until Blair fell in on top of me and yelled that there was an ambush and I knew I was going to die. Worse, be captured, tortured like they said you were tortured, made to run through the jungle without your boots and be cut with knives and beaten, with split bamboo. Peterson was just behind, then your father . . . never thought the fucking rotors would start; thought the engine had been hit by a shell, broken. Then they fired. All the time I was looking out at those who were trapped, expecting them to come on: that's why I started the engines at first, wanting to be ready as soon as they were aboard. But they didn't move and I realised they were pinned down and weren't going to come . . . Blair was on the gun, the starboard gun, but he wasn't used to it and it jammed almost at once and he tried the M-16 but that wasn't any good because there were too many of them. Peterson was screaming . . . your father too . . . so frightened; everyone was so frightened. I know it was Peterson who shouted "get out" although he said it wasn't, later. Didn't matter, not really, because I already had the blades pitched for lift: didn't want to be caught and tortured. Fucking stupid war was over anyway. Got to about twenty feet and we could see perfectly what was happening, down below. Forest had them into a perfect defensive position, like I told you earlier: Vine in the middle, with the kids, everyone else around them. Actually saw Vine fall. Some kids, with him: one of the soldiers, too, although I didn't know his name . . .' Patton's crying worsened, sobs racking through his body. 'Christ they were brave, so brave,' he resumed. 'Blair had the other gun going by now, keeping the 'Cong back. I went towards where they were held down and Peterson yelled at me, to get out. Down below Bartel thought we were coming back for them. I saw him signal, to a smoother spot in the scrub where I could land. Except that I didn't land. I kept climbing and when he saw what was happening Blair stopped firing and said we should go down but Peterson said no and I didn't intend going down anyway . . . we could see everything, where we were. The Vietcong realised we weren't coming down and there wasn't any danger from the sidegun and they began easing out from the protection of the thicker trees. Forest and Bartel weren't firing, not at my last sight of them: they knew we weren't coming back and just stood there, looking up . . . only McCloud reacted. He tried to shoot at us, with a handgun, but it was stupid because we were too high.'

He had it! thought Hawkins, triumphantly. He had the truth that could break and destroy a President before he assumed office. What else would be broken and destroyed with it? It didn't matter. Only one thing mattered. Himself and Eleanor.

'The helicopter was hit, like I said,' took up Patton. 'I managed to clear the trees but only just. Peterson recovered quicker than anybody else. When Blair said he'd file a report Peterson replied that he was under military jurisdiction: that the responsibility wasn't his. I was flying the helicopter so I couldn't see but I could hear because the noise was such with the sides open that everyone had to wear helmets and use the open system. I knew, without even looking, that Blair had been as frightened as anyone and was as relieved as anyone to be out. Peterson recognised it too. Peterson was an important guy, even then: like I said, a lot of clout. We all knew that. He promised Blair he'd look after him: make sure he got the right recognition. Me, too. That's how the business got started, after Nam: not just the money but how all the contracts came in . . . contracts that the bastard Harriman blocked, although I can't prove it, after I told Peterson what we'd found, when we rescued them.'

The switch was too quick. Not as concerned with breaking the flow, now that he knew, Hawkins said, 'What about my father? You haven't told me about my father?'

'Do you want me to tell you about your father?' said Patton, hands away from his face for the first time for over an hour.

'Yes,' said Hawkins.

'Your father was a jelly: out of his mind. We'd decided everything before he got his head together and when Peterson said how we intended it to be he just said okay, whatever we wanted . . .'

Hawkins swallowed, physically holding down the sensation of sickness. What else would he break and destroy, he thought again; *had* broken and destroyed. 'It couldn't work,' he insisted. 'You *knew* they were alive when you left them: you *saw* them alive.'

'I told you before!' said Patton, with something approaching exasperation. 'The helicopter took eighty-five hits: all the glass was gone, shattered. None of us ever said they were dead: we said we saw them shot down and we *presumed* they were dead and the condition of the machine made it difficult for us to see anything, with a hundred per cent accuracy. But sure we worried: those first weeks and months we worried more than you'll ever know. Because when we created the story, we never realised – not even your father realised – the effect it would have. It made us, all of us. I got the business contracts' – Patton gave a faint, sad smile – 'Peterson promised them back when he got into power but it won't happen now, will it? Blair got his promotions. Your father got his recognition. And look what Peterson's got . . . All of us with so much to lose . . .'

Hawkins could still taste the sickness, for a different reason now. 'What happened when Ninh and Nicole came out?' he said, determined upon a full, absolutely destructive exposé.

'It was like flying back to Saigon, all over again,' said Patton. 'We met Peterson in Washington, the day after Jordan came to me for a boat and Blair was told he was in charge of the rescue. Peterson said it could blow everything: not just his election but Blair's career and my business too. We had to work together, like we'd worked together before.'

'How?' pressed Hawkins, wanting a tramlined narrative now.

'He said I was to go aboard the freighter, so that I could keep him in constant touch. And that Blair was to kill them, if he found them alive . . .'

'*Kill* them!'

'There'd been some training already, at Camp Peary: briefings at least. A lot about psychology and the sort of condition they'd be in, if they had survived. There'd been talk of sedation and Peterson said Blair should over-sedate and that it could be explained away that in the final moments of rescue it had been too much for them and their bodies had over-reacted.'

'He was going to do *that*!'

'He said no, that he wouldn't. That if they were there he'd bring them out and face whatever enquiry there was. And that's what he was doing: bringing them out when he was killed.'

'That's why you cabled Peterson, from the ship?' reflected Hawkins.

'I didn't think he'd make the announcement that he did: I meant what I told you on the ship. We didn't know how it was then, how badly they were affected, but Blair's radio message said it looked bad and he phrased it in the way we'd rehearsed, on the trip from Manila, so that I would understand just *how* bad it was; another decision from the Washington meeting with Peterson was that if they did get out, with some mental problems, and if they ever accused us of abandoning them we could say it wasn't true but simply a psychosis, something that would be explainable after such long imprisonment.'

'*Jesus*!' said Hawkins, unable any longer to hold in the disgust.

'Don't sit in judgment on me!' said Patton, suddenly angry. 'Don't sit in judgment until you've really been frightened . . . until you really know what it's like . . . until you know that you're a coward . . . and live with it . . .'

'I'm going to sit in judgment upon you,' declared Hawkins, his mind made up. 'I'm going to sit in judgment upon you and upon my father and upon Blair and most of all upon Peterson. And when I've finished there won't be a hole deep enough for you and Peterson to hide in!'

Won! he thought.

The euphoria became more rational on the shuttle back to Washington, when he put the pieces together and made his picture and realised how it would appear, to anyone looking in from the outside. It meant the book would have to be scrapped; but then it should be, because it was an untrue book. That his father would be posthumously stripped of his fame and his honour and his reputation too; shown instead to be a coward and a cheat and a charlatan. But that should happen as well, because it was the truth, at the end at least, and he intended telling the truth. Patton, out of his own mouth, would be destroyed. Deservedly. And so would Peterson. Like a child hoarding the red sweet with the cherry-filled centre, Hawkins put aside until last his consideration of the effect upon Peterson, savouring the taste and the flavour. Hawkins knew he'd already established a reputation; but this would elevate him to a level where he could look either side and see no one, not for miles, no peer or competitor or challenger. Unique. He felt again the sensual but long-ago tingling feeling, like that moment when the whisky he no longer needed or thought about first washed over the barriers, to pick up his senses and carry them along in the tideswell.

Eleanor was wrong to be so frightened: the very fame and recognition he would achieve, from doing what he was going to do, would be their protection. Sure it would be life inside the goldfish bowl, for a while: but he – and through him, Eleanor – would have the stature to stare down anyone ogling in. It was understandable that Eleanor was bewildered and overwhelmed by what had happened, knowing as he did how much she hated and despised the exposure to which she was subjected now and feared in the future. She just needed reassurance: a proper explanation and a proper reassurance. Hawkins decided he felt complete: in control of himself and in control of his life and in control of everything. Complete.

Hawkins couldn't remember the last time he'd had a drink but he went into the bar at National Airport and ordered a Scotch and stared down into it, unable to comprehend – even though he knew the reasons well enough now – why he had needed its buoyancy for so long. Supporting him while all the time sinking him, he thought. Thank Christ he'd learned to swim. He raised the glass finally, in a solitary, unspoken toast to himself: the whisky tasted harsh, burning his throat and he didn't bother to finish it, not wanting to.

Ninh and Nicole were waiting for him in Maryland Avenue and while they started to prepare the evening meal Hawkins wondered about the effect upon them of what he intended to do. Nicole knew a lot already: Ninh probably guessed some of it, if the man had been that close to his father. Still shocked, though. Hawkins decided he'd have to explain but not yet: nearer the time of publication.

It was still only minutes after his return when the bell sounded and because the refugees were both in the kitchen Hawkins answered it, staring in surprise at Rampallie. The timing meant the man must have been waiting for him: Hawkins looked beyond Peterson's campaign manager but couldn't see a car. No one else either.

'I'd like to talk to you, Ray,' said Rampallie, eyes quivering behind the spectacles.

Hawkins hesitated, then stood aside and said, 'Come in.'

The American remained enquiringly in the hallway and Hawkins gestured towards the study. Inside Rampallie didn't sit, standing instead before the fireplace and looking around, particularly at the typewriter on the cluttered desk.

'Gather you've been on a trip, to New York,' said Rampallie.

'Patton keeps in close touch, doesn't he?' said Hawkins. He supposed it was obvious the man should have contacted Peterson but he hadn't expected it.

'I don't know the details,' said Rampallie. 'I've just been asked to come and say the President wants to see you.'

'Is he that, yet?' said Hawkins.

'The election's over,' insisted Rampallie. 'He's the President.'

'Do you know what it's about?'

'Like I said,' replied the American. 'I don't know the details. I was just told to mention New York and ask you not to do anything unwise until after you've spoken to the President.'

'Was that the word, "unwise"?' said Hawkins.

'Yes,' said Rampallie. 'Unwise.'

Hawkins remembered the first meeting with Nicole aboard the freighter off Hong Kong and his awareness of how words never properly matched occasions in their importance: it was happening again, now. Hawkins decided he had nothing to lose from any encounter: everything to gain, in fact, because he would refuse to be bound by any understanding of confidentiality and actually get Peterson's reaction to the accusations. 'When?' he said.

'The President is going to be at home in Georgetown, tomorrow,' said Rampallie. 'He suggested the afternoon, around three.'

They could have been arranging a casual social occasion, nothing more important than tea or cocktails, thought Hawkins. 'Three will be fine,' he said.

'We appreciate it,' said Rampallie, politely.

'I'm not sure you will,' said Hawkins.

It was a sudden decision of Hawkins to detour, after he left Maryland Avenue for his office and go first instead to his bank and hire the safe deposit box for the tape of the previous day's meeting with Patton. It was a straightforward transaction but he was still later arriving than he normally liked.

He took two wire services, UPI for international, and Associated Press for domestic, within America. The paper was ribboned across the floor when he got into his room. He tore it off from the machine first, then sectioned it to make reading easier, going quickly through to find anything worth expansion for Sunday.

The item about Patton was on the third slip and was slugged 'war hero', leading off with the Chau Phu incident and linked with the Can Tho rescue, eight years later. The details were in the second paragraph. Patton had been found in the Third Avenue headquarters of his multi-million dollar conglomerate by office cleaners and preliminary medical examination indicated he had been dead for several hours, possibly from the previous evening. Empty pill containers had been taken away for examination and an autopsy was scheduled later in the day. Although a number of government contracts had recently been cancelled, a company spokesman said there were no financial difficulties. There was no note or indication why the tragedy had occurred. 'A mystery' was how the spokesman had described it.

Chapter Thirty-Two

That part of Dumbarton Avenue upon which Peterson lived was partitioned by crush barriers. There were police cars and unmarked vehicles forming additional protection lines behind them and Hawkins recognised as he approached that any transitional period between an election and an inauguration must be a nightmare for the Secret Service entrusted with the security of a President with a town house like this. There were check points at each end of the cordoned street, manned by plain clothes men with ear-piece receivers and walkie-talkie handsets. Hawkins used his White House ID and his appointment was checked off against a clipboard and a man allocated to escort him to the door which was opened by another plain clothes agent. The man apologised for the necessity and ran his hands over Hawkins' body and stopped at the bump of the tape recorder in his jacket pocket, asking him to produce it. Hawkins did and the man checked that it really was a recorder and returned it. A third man led Hawkins up into the book-lined drawing room on the first floor, with the view of the patio, in which he'd drank and talked with Eleanor all those months ago. He'd returned that time from New York after a conversation with Patton, Hawkins remembered. He wondered if Eleanor was in the house: whether he would see her.

Peterson must have been waiting. He followed almost immediately into the room, thanked the Secret Service escort dismissively and said 'Hello Ray', without any offer of a handshake or greeting to suggest they were friends, for which Hawkins was grateful because any artificiality between them was over and they both knew it.

Now that the encounter was happening, Hawkins was unsure how to conduct it: inadequate words, he thought again. He said, 'I should have guessed Patton would have called you, but I didn't.'

'Why don't we go out on to the patio?' said Peterson. 'It's cool but I'd rather be in the open.' The President-elect led the way, sliding back the windows but then standing back, for Hawkins to precede him out of the house. Hawkins sat in the chair that Peterson indicated, on the far side of the table, watching while the American carefully closed the windows behind them.

'You've got a tape recorder in your left-hand pocket,' said Peterson.

The protection was extremely efficient, Hawkins thought. 'Yes,' he said.

'I don't think we want that, do we Ray?'

'Records no longer important as they once were?'

'Let's try to keep this on a civilised level, shall we?'

'I would have thought that might be difficult.'

'The recorder's still in your pocket.'

Hawkins placed it on the table between them and Peterson ensured it was off and said, 'Thanks.'

'There's still proof of our meeting,' said Hawkins, recalling the clipboard check-list in the street outside.

'Have you heard about Patton?' asked Peterson, ignoring the remark.

'Yes,' said Hawkins. Peterson was controlling very well the nervousness he had to be feeling.

The American shook his head in an expression of sorrow and said, 'A tragedy committing suicide like that.'

Hawkins despised the man for the hypocrisy. He said, 'Is that what it was, suicide?'

Peterson frowned, head to one side. 'What else?'

'You tell me.'

'Ray! Come on! What sort of suggestion is that?'

'It wasn't a suggestion, it was a question.'

'Which I don't understand,' said Peterson, who did.

'He told me about it; everything,' said Hawkins. 'I know everything.'

'He called me,' reminded Peterson. 'He was very upset.'

'You going to tell that to the authorities in New York?'

Refusing that remark too, Peterson said, 'Gives you quite a story, I guess?'

'Good enough to keep you out of the White House.'

Peterson winced, an exaggerated expression, and said, 'That's a pretty extreme threat.'

'It's not a threat,' said Hawkins. 'It's a fact.'

'You're not troubled, having Patton's death on your conscience?'

'Are you, for the deaths of those who died at Chau Phu when you ran away? Or for the three who survived and are basket cases? And always will be.'

'You ever been frightened, Ray? Not ordinary frightened: everyone thinks they've been that. I mean *really* frightened, so frightened that you can't think, don't think. So frightened that afterwards, directly afterwards, you can't remember what you've done or what you said or even where you are. Just bits. Like I can easily remember worrying that your father couldn't keep up, because he was too old and too out of condition: that the kids were so sick they might not survive. And that initial terror when I realised we were under fire and that I might get hurt. Die. I remember praying, begging God not to let me get hit, promising that I'd do anything if He spared me any hurt. You ever been *that* frightened?'

'Did you write that or was it one of your speechwriters?' demanded Hawkins, unimpressed.

Peterson's face visibly closed against him. 'OK,' said the man, the tone hardening. 'What is it? Set it out, what you want.'

Something you can't give me: or wouldn't give me, not willingly, thought Hawkins. He said, 'Nothing: nothing at all. Just to see you exposed, for what you did. What you are.'

'You know what happens, during a period like this, between an election and an inauguration?' said Peterson. 'There's a cooperation, even though it's differing parties. My people get eased in, to take over from Harriman's on the way out. Ensures a smoothness . . .'

'You making a point?' said Hawkins.

'Already found something interesting about the Vietnamese and the woman: something you already know, I guess. About their entry into this country. Harriman was a cautious son-of-a-bitch: only let them in on Presidential order.'

'I know that,' said Hawkins, feeling a stir of unease.

'That's the way it stays,' said Peterson. 'Until the President decides otherwise, they're here at his descretion, his say-so.'

Understanding the threat, Hawkins said 'To revoke that order – to make them stateless – you've got to take office as President. You think you stand a chance of that?'

'I could do it,' said Peterson. 'Even if I was driven from office, disgraced, there would be a lot of people who would stay behind: wouldn't like what happened to me, because of you. I'd make sure they were expelled . . .' he allowed the gap. 'Or alternatively given legal permanency, if I felt sufficient gratitude.'

'I could get them into England easily enough,' said Hawkins, enjoying the power of being able to reject the bribe.

'You want to bet?'

'Do you?' said Hawkins.

'Do you think you could stay on here in America, whatever happened?'

'I wouldn't want to,' said Hawkins. Because Eleanor wouldn't.

'You must want it pretty much – whatever it is you want – if you're prepared to destroy your father.'

'I'm not setting out to destroy my father,' said Hawkins. 'I'll tell the story of an old man who just didn't know when to stop trying. Who wanted to be brave when it wasn't possible any more.'

'I've been very reasonable: very civilised,' said Peterson. 'I don't think you've really any idea what you're thinking of doing. How many people might be affected, affected enough to want to stop you?'

'Like Patton was stopped?'

'Don't be stupid,' said Peterson. 'Patton committed suicide, pure and simple. And with him went what he told you. You're going to write what you were told by a deranged man, you know? Are your publishers and your editors going to be satisfied, printing the ramblings of some mentally sick guy against the denials that we can put up? We know nothing's happened at the clinic, incidentally. We've checked, just to make sure.'

'I made sure, too,' said Hawkins, still confident. He leaned forward, tapping the dead tape recorder. 'It wasn't stopped when I spoke to Patton,' he said. 'It was running and it got every word. Not the words of a mentally deranged man but the words of someone glad at last to make an admission, to tell exactly what happened and who was responsible for making it happen. My publishers and my editors are going to be satisfied enough to print, when they hear it.'

'You're going to cause yourself a lot of trouble, Ray. More trouble than you ever thought possible to befall anyone.'

'Go to hell,' said Hawkins.

When Hawkins got back to Maryland Avenue they both ran from the house to meet him. Nicole was crying too much to be able to explain and Ninh was shaking too, the words jerking and disjointed, and it wasn't until he got inside the house that he saw the reason for their distress. It had been a methodical search and very professional. Every room had been scoured, particularly the basement where the carefully annotated contents of the boxes were heaped and scattered in hopeless, irreparable confusion. The basement was the worst but every other room was wrecked and when he tried, anticipating the result, Hawkins found that every sound and video tape he possessed had been wiped clean by some electronic device. A magnet would have been sufficient, he supposed.

Hawkins' calmness calmed the other two and Ninh said, 'The police. We should get the police.'

Hawkins shook his head, knowing their fear of officialdom and of uniforms, and knowing too, the pointlessness. 'No,' he said. 'The police won't find anything.'

Hawkins couldn't find his address book in the mess, but he didn't need it anyway, the number of the Georgetown house memorised from a hundred surreptitious attempts to contact Eleanor. It still took several minutes for him to be connected to Rampallie and when he was Hawkins said, 'I'd like you to give a message to . . .' he paused, then completed '. . . to the President. Tell him you missed it.'

Chapter Thirty-Three

Eleanor called him at the office, which she'd never done before, when he was trying to compose the message to London suggesting he should personally return to explain why the unveiling had to be postponed and the book rewritten so completely that publication would have to be delayed, possibly for several months. Contacting Eleanor had been his other preoccupation and he agreed at once to the meeting, because Eleanor was his predominant consideration; contacting London could wait until the following day. It had to be another aimless car ride, the only way any encounter was safe for them and that uncertain in the daytime; like the last occasion she went out to the peripheral beltway, a circular journey to nowhere.

Hawkins was talking long before they reached it, babbling the words, anxious for her to understand that everything was going to work as they wanted and that there was a way to be together, always.

'It's not going to be easy,' he said. 'But then we never expected that it would be. In fact it's going to be hell; I won't lie to you about that. But there's a way and it's going to work. I'm . . .'

'I know,' she said, cutting him off.

'What?' he frowned.

'I said I know.'

Hawkins smiled, at her misunderstanding. 'No, darling,' he said. 'Let me explain what's going to happen . . .'

'John told me,' she interrupted again. 'Confessed everything, last night. Cried, like a child and said he was sorry for all the hurt and harm he was going to cause me and that I was going to have to undergo the sort of exposure I couldn't conceive . . . imagine even. And then he told me. All about the mission and how they ran and how they thought they would be exposed and how terrified he was, when Ninh and Nicole got out and then when the survivors were rescued . . .'

Hawkins was sitting twisted in his seat, looking directly across the vehicle at her. 'He can't have told you everything,' he insisted, hurriedly. 'Not what he asked Blair to do . . .'

'Everything,' insisted Eleanor. She got to the beltway and turned south.

He reached across, wanting physical contact with her, squeezing her arm. 'Then you see how it's going to work,' he said. 'Let me tell you what I'm going to do . . .'

'No,' she refused. 'I don't want to hear it.'

'But darling, you've got to. If you don't let me explain you won't fully understand.'

'Nothing's changed,' she insisted.

Hawkins blinked across the car at her, trying to understand. 'What hasn't changed?'

'Us,' she said. 'I still can't go through with it.'

'It's the *way*,' he said, shouting and not meaning to. 'The way for us to be together.'

'I couldn't stand it,' said Eleanor. 'It's bad enough now and it would have been bad enough if we'd just tried to get a divorce but it would be terrifying this way. I couldn't stand it; just couldn't.' She shivered, as if she were physically cold.

Consciously controlling himself, Hawkins said, 'Darling, now listen to me; please listen. I'd protect you; look after you. I told you it would be hell and so it would, but not for ever. We could get out of America, whenever you liked: go to England. You know what it's like there: how easy it would be to settle there . . . You would even get custody of the children this way.'

'No!' she said, positively. 'I meant it last time and I mean it even more now. I can't. Won't.'

'So you're going to go on living with him!'

'No,' she said at once. 'Not like that anyway. We'll occupy the same houses and appear in public in the same events but when the door's closed, that'll be it. That's the way it's been for a long time anyway, so nothing will be different.'

'So you'd rather live a sham like that than be happy with me?'

'I couldn't be happy with you,' she said. 'I told you last time that before we had time to settle down into the sort of existence you're promising, anything between us would be dead. Finished. And if you're honest with yourself, you'll acknowledge it too.'

'I *love* you.'

'I love you,' she said. 'And I want to go on loving you. Which is why I'm saying no.'

'That's ludicrous.'

'I know all the words to describe it,' said Eleanor. 'I've said them to myself over and over again.'

Ahead the indicators appeared, for the southern turn-offs. Eleanor kept to the beltway. They drove in silence for several miles and then she said, abruptly, 'Don't!'

'Don't what?'

'Expose him.'

Hawkins snorted a disbelieving laugh. 'You're trying to protect him!'

She laughed back, just as humourlessly. 'Dear God, no! I loathe and despise him, for the person he is and always has been and always will be. I'm asking you for myself, just for me. If you destroy him, you'll destroy me. I'll be caught up in it. The children too. It'll be inevitable . . .' Another shudder. 'Horrible!'

'You can't ask me not to do it,' said Hawkins.

'I can,' she said. 'I love you and because I love you I can.'

'No!' he said.

She shrugged, a defeated, resigned gesture. 'I had to ask,' she said. 'I believed you loved me so I had to ask.'

'I do love you!'

'Then let me go on living,' she said. 'Living as much as I do, at least. I know everything is false-fronted and a sham and I know it's stupid but let me keep a false front to hide behind at least.'

'You know what you're asking me to do?' he said, anguished.

'Maybe not,' she admitted. 'Maybe not completely.'

'Let me do it my way,' he pleaded desperately. 'It would work; I know it would work.'

'No,' she said.

Unlike Eleanor – who had taken nothing – Peterson had drunk a lot of wine at the Inaugural Ball and was now drinking brandy and was astride the actual point of drunkenness, at the moment of full confession. 'I was sure I'd covered everything,' he said. 'Everything! I got Blair the decorations and the promotions and I funded Patton and got him the government contracts and because of the promises I persuaded that stupid, stumbling old son-of-a-bitch Hawkins to get himself posted here to Washington and fed him so much stuff that he became the best commentator there was; and took everything about Ninh and the woman from him because I thought I was closing the gap by saying I'd get them out of Vietnam. By actually *making* the application and sending it with the names and documentation to Hanoi – the very thing he wanted – I thought I'd guarantee they'd never get out. Be in jail forever.'

Eleanor looked at him despisingly and said, 'You've already told me.'

He looked at the discarded newspapers around the chair where she sat and said, 'Camelot! That's what they're calling it! Camelot all over again.'

'I read that, too.'

'Is this how it's going to be?'

'How what's going to be?'

'Us. Hostile?'

'We don't have to spend too much time together,' she said. 'Only the bare minimum.'

Peterson poured himself more brandy and said, 'I knew how it was, from the beginning, an arrangement with your father, so I could get the right start, but I'd always hoped it would be better.'

'You've got very good at it,' said Eleanor. 'Expert.'

He squinted at her over the rim of his brandy bowl. 'What are you talking about?'

'Lying sincerely,' she said. 'You never gave a damn how it would be. You wanted me because of the help my father could give you and you used John junior when your fucking ratings slumped – I'll never forgive you for that – and then when you thought you were in trouble, you used me.'

'I'd never have used the Secret Service reports, not unless I'd had to. Had no reason for letting you know how early on you were under surveillance. It was for your own protection.'

'Bullshit!' she said.

'Did you love him?'

Eleanor didn't reply at once. Then she said, 'Oh yes. Yes, I loved him. Still do. He's gentle and kind and he was a good lover . . . having a good lover made a pleasant change. It would have been wonderful, being married to him.'

'But this is better?' said Peterson, gesturing around the private White House apartments.

'I'll try to make it bearable,' she said. There was another pause and then she said, 'You would have done it, wouldn't you? Destroyed him, I mean, if he'd published. You'd have done something to bring him down.'

'Of course,' said Peterson, as if he were surprised at the question. 'And I don't want you to try to retain any contact.'

'I won't,' she promised.

'I'll always be grateful,' said Peterson.

'Oh, you will,' she said, this time her turn to show apparent surprise. 'For what I did for you I'm going to make you be grateful to me every day for the rest of your life.'

He grimaced at her, not taking the threat seriously. 'America's first lady President,' he said, raising his drink in mock toast.

'Yes,' she said. 'Just that. And don't you ever forget it, for a moment.'

Peterson wanted to go to bed, to a bedroom separate from hers, but Eleanor made him walk with her through the Presidential home and as they entered the East Wing lobby Peterson asked, 'What do you want to come here for?'

Eleanor walked slowly up and down the corridor, gazing at the portraits of previous First Ladies, finally returning to where he stood, at the entrance to the Garden Room.

'There!' she announced, pointing to an empty space on the wall to the right. 'That's where I want my portrait to be.'

All the speeches referred to his father as an outstanding journalist and a man of honesty and integrity and after the unveiling ceremony there was a reception in the mahogany panelled boardroom where gift copies of the book were arranged on a central table, personally signed by Hawkins.

'Read it already,' said Lord Doondale. 'Damned good.'

'Thank you,' said Hawkins.

'Going to make a good serialisation.'
'I hope so,' said Hawkins.
'Gather you want to leave Washington?'
'I think it's time to move on,' said Hawkins.
'Be difficult to fill the gap there,' said the proprietor. 'Established quite a reputation for yourself.'
'I think Harry Jones wants it.'
'Thought where you'd like to go?'
Hawkins shook his head. 'We've decided to work that out when I get back here.'
'Whatever you want,' said Doondale. 'Meant what I said that day at lunch.'
Hawkins limited himself at the reception, knowing it was important to do so, and afterwards didn't go into Fleet Street because he didn't want to encounter anyone he knew. Instead he walked up Ludgate Hill in the direction of the City and found a bar just before St Paul's Cathedral.
'Whisky,' he said to the barman. He widened the gap between his thumb and forefinger and added, 'Large.'

Epilogue

The influx of Indochinese refugees into this country over the past few years has resulted in convincing evidence that the Vietnamese and Lao continue to hold Americans in captivity. The Defense Intelligence Agency (DIA) is currently investigating over 400 eyewitness reports pertaining to men still held captive: some sightings are as recent as December, 1981.

As a result, US government policies and attitudes are changing. Where for many years official statements claimed a lack of credible information that men were still alive, new government policy reflects the increasing evidence that some Americans are still in captivity. According to Lt. General Eugene Tighe, former director of the DIA, all US intelligence collection disciplines are now being brought to bear on locating POWs. In June, 1981, General Tighe told a congressional subcommittee that in his personal judgment, Americans are currently held captive in Indochina.

Statement of the Washington-based National League of POW/MIA Families

Your long vigil is over. Your government is attentive and the intelligence assets of the United States are fully focussed on this issue . . . I call upon the government in Hanoi to honour their pledge to the American people on the prisoner of war and missing in action issue. Not for me, not for our government but for our missing men and those of you who did nothing to deserve this terrible ordeal.

United States President Ronald Reagan at a meeting of that League, 28 January, 1983.